THE ART OF GOLANG

A BEGINNER'S GUIDE

APURV V BRAHMBHATT

BlueRose ONE .com
Stories Matter
NewDelhi • London

BLUEROSE PUBLISHERS
India | U.K.

BlueRose ONE.com
Stories Matter
NewDelhi • London

For permissions requests or inquiries regarding this publication,
please contact:

BLUEROSE PUBLISHERS
www.BlueRoseONE.com
info@bluerosepublishers.com
+91 8882 898 898
+4407342408967

ISBN: 978-93-6783-463-3

Cover Design: Shubham
Typesetting: Sagar

First Edition: January 2025

Contents

Chapter 1. Introduction

1.1. Overview

The software development industry is constantly evolving and is home to a wide variety of programming languages. Python, Node.js, and Java are widely used by developers and in recent years, the use of Golang has skyrocketed in the back-end development sector, making it the most sought-after skills in the industry. These programming languages have proven to be reliable, efficient, and versatile, making them ideal for a wide range of projects and applications. Node.js, Java, and Python are popular programming languages used in the software development industry today. Each of these languages has its own strengths and weaknesses, making them well-suited for different types of applications and use cases.

Node.js is known for its fast performance, making it an ideal choice for real-time applications such as chat apps, online games, and streaming services. Node.js is built on JavaScript, which is one of the most popular and widely used programming languages, making it easy for web developers to pick up and use. Additionally, Node.js has a large and active community of developers who contribute to its ecosystem by developing and maintaining a wide range of packages and modules.

Java is a popular choice for building enterprise-level applications due to its scalability, reliability, and security. Java has been around for over two decades and is used by many large corporations, including Google and Amazon. Java has a vast library of libraries and tools, making it easy to build complex applications. Java is also designed to be portable, meaning that applications can be run on any platform that supports Java.

Python is a popular choice for data science, artificial intelligence, and scientific computing applications. Python is known for its simplicity, readability, and ease of use. This makes it a popular choice for beginners, as well as experienced developers who want to quickly prototype and build applications. Python has a vast library of packages and modules, making it easy to integrate with other tools and technologies.

Each of these programming languages is popular for different reasons and has its own strengths and weaknesses. The choice of language for a particular project will depend on the specific requirements and goals of the project, as well as the skills and expertise of the development team.

Whether you are a seasoned developer or just starting out, this book will provide you with a comprehensive understanding of the Go language and its integration with various databases such as SQL and NoSQL. In this book, we will delve into the intricacies of using Golang with these databases, exploring the strengths and weaknesses of each type of database, and how they can be used to create efficient and effective back-end solutions. This book is your blueprint to success in the world of Golang and databases.

History

Go, also known as Golang, is a programming language created by Google in 2007. The language was developed by a team of engineers at Google, including Rob Pike, Ken Thompson, and Robert Griesemer, with the goal of creating a fast, efficient, and reliable programming language for large-scale web applications and distributed systems.

Go was born out of the need for a more scalable, efficient, and reliable alternative to existing programming languages. At the time, Google was dealing with complex and resource-intensive web applications and distributed systems, and they found that traditional programming languages, such as C++, were not well suited to these types of systems.

The Go project officially began in September 2007 and was initially released as an open-source project in November 2009. The language quickly gained popularity among developers, who appreciated its simplicity, performance, and efficiency.

One of the lesser-known facts about Go is that it was originally designed as an experimental language, but its popularity and success have made it a staple in modern software development. Another interesting fact is that Go is often referred to as a "systems programming language," which means that it is designed for use in low-level systems programming and not for general-purpose applications. In addition to its use in web applications and distributed systems, Go is also widely used for network programming, game development, and in the development of machine learning and artificial intelligence applications.

Despite its relatively short history, Go has already made a significant impact on the software development community and continues to evolve and grow. With its simple syntax, efficient performance, and focus on scalability, it is clear that Go will continue to play a crucial role in modern software development for years to come.

Key Features

Go is designed to be a language that is both efficient and easy to use. Here are some of the key features of Go that make it an attractive option for developers:

1. **Concurrency**: Go is built with concurrency in mind, making it possible to write programs that can run multiple tasks simultaneously improving performance and efficiency. The language provides a built-in concurrency model using Goroutines and Channels, which makes it easy for developers to write concurrent programs without having to worry about deadlocks, race conditions, and other common concurrency problems. This makes Go a great choice for developing large-scale, high-performance systems that need to handle a large number of concurrent requests.

2. **Garbage Collection**: Go has a built-in garbage collector that automatically frees up memory that is no longer needed. This feature also greatly reduces the time and effort required for manual memory management, freeing developers to focus on other important aspects of their code. The garbage collector also runs in a concurrent manner, ensuring that its impact on performance is minimal. The automatic memory management provided by

the garbage collector helps to make Go a highly efficient and reliable language for building scalable applications.

3. **Fast Compilation**: Go has a fast compiler, which means that programs can be compiled and run quickly. This makes it easy to develop and test programs, even for large projects. The fast compilation time is achieved by a combination of efficient algorithms, smart data structures, and parallel processing. This makes Go a great choice for building high-performance systems, where quick turnaround times are essential.
4. **Static Typing**: Go is a statically typed language, which means that the types of variables are known at compile-time.This means that the compiler can catch type-related errors early on in the development process, making it easier to debug and maintain the code. Additionally, the use of static typing allows for better optimization by the compiler, leading to faster execution and improved performance. This also provides a level of safety and security that is not found in dynamically typed languages.
5. **Cross-platform Support**: Go programs can be compiled for a variety of platforms, including Windows, Linux, and macOS. This helps to increase the reach and scalability of Go-based applications, making it a versatile language for modern software development. Furthermore, Go's binary-only distribution model ensures that applications can be deployed without the need to install a runtime or any other dependencies, making it a convenient choice for deployment in cloud environments and other constrained environments.
6. **Simple Syntax**: Go has a simple and straightforward syntax that is easy to learn and use. This makes it a great option for developers who are just starting out, as well as for experienced developers who are looking for a more concise and efficient language.
7. **Standard Library**: Go has a comprehensive standard library that includes many commonly used functions and packages. This makes it possible to write programs quickly and efficiently, without having to rely on third-party libraries.

8. **Robust Community**: Go has a large and active community of developers who are constantly contributing to the language and developing new packages and libraries. This makes it possible to get help and support when needed, and to find solutions to common problems.

1.2. Getting started with Go

You need to set up a development environment on your computer which typically involves installing the Go programming language and an Integrated Development Environment (IDE). Setting up a proper development environment is important as it allows you to write, test, and debug your Go code in an efficient and streamlined manner. A good IDE will provide features such as syntax highlighting, code completion, and error checking, making it easier for you to write and maintain your code. Additionally, having a well-configured development environment will allow you to quickly navigate your code, and make it easier to collaborate with other developers on your projects.

Installing Go

To install Golang, follow these steps:

1. Go to the official Golang website at https://golang.org/.
2. Click the "Downloads" link at the top of the page.
3. Choose the appropriate download package for your operating system.
4. Run the installer and follow the prompts.

Alternatively, if you are using a Linux or Unix-based system, you can install Golang using the package manager. For example, on Ubuntu or Debian, you can use the following command:

```
sudo apt-get install golang
```

After the installation is complete, you can check that Golang is installed correctly by opening a terminal window and typing the following command:

```
go version
```

This should output the version number of Golang that you have installed.

Configuring the Environment

Once you have installed Golang, you need to configure your environment. This includes setting the environment variables and configuring the editor or IDE you plan to use.

Setting the GOPATH Environment Variable

The GOPATH environment variable is the root directory for your Golang workspace. This is where your source code, libraries, and other files will be stored.

To set the GOPATH variable, follow these steps:

1. Create a directory for your workspace, for example, ~/go.
2. Set the GOPATH environment variable to the path of your workspace directory. On Unix or Linux-based systems, you can do this by adding the following line to your shell configuration file (e.g., ~/.bashrc or ~/.bash_profile):

```
export GOPATH=~/go
```

On Windows, you can set the environment variable through the Control Panel.

Setting Up IDE

Some popular options for IDEs are Visual Studio Code, GoLand, Sublime Text, Atom and Eclipse. It's important to choose an IDE that meets your needs and preferences, as it will help streamline your development process and improve your productivity. The examples provided in this book can be easily copied and pasted into any of the listed IDEs for quick testing and experimentation

The setup instructions for each IDE can vary greatly, and it is beyond the scope of this book to provide detailed steps for each one. However, each IDE typically has its own website with comprehensive documentation on how to set it up. For the best experience, we recommend referring to the specific website of the chosen IDE for setup instructions.

Chapter 2.
Basic Concepts

2.1. Variables

A variable is essentially a name given to a storage space in a computer's memory where you can keep data. When you create a variable, you assign it a specific type of data it can hold, like a number or a piece of text. In Golang, variables are strictly typed, meaning you must declare the type of data a variable will hold when you create it. This strict typing helps ensure that your program uses data correctly, reducing errors and enhancing the program's reliability. During the execution of the program, you can change the value stored in a variable as needed.

Declaring variable

In Go, declaring a variable is the process of defining a name and a type for a piece of data that you intend to use in your program. This step is essential as it tells the Go compiler what kind of data the variable will hold and allocates the appropriate amount of memory for it. You declare a variable using the var keyword, followed by the variable name and its type. Here is how it looks in practice:

```
var age int // declaring an integer variable named age
```

We can also initialize a variable at the time of declaration like this:

```
var name string = "John"
```

Golang also provides a shorthand notation to declare and initialize a variable at the same time like this:

```
age := 30 // shorthand notation for declaring and initializing an integer variable
```

It is called type inference, where the type of the variable is automatically inferred by the compiler based on the value assigned to it. For example:

```
name := "John"
```

In this example, we declare a variable name and initialize it with a value of "**John**". Since we did not specify the type, Golang infers the type to be a string.

Declaring multiple variables

It is possible to declare multiple variables of different data types using a single var clause in Golang. To declare multiple variables of different data types, we can use the following syntax:

```
var (
  name string = "John"
  age int = 30
  salary float64 = 2500.50
)
```

In the above example, we are declaring three variables, name of type string, age of type int, and salary of type float64. Each variable is assigned a value in the same line where it is declared.

We can also declare multiple variables without specifying the data type, and the compiler will automatically infer the type based on the value assigned. For example:

```
var (
  name = "John"
  age = 30
  salary = 2500.50
)
```

In this example, the compiler infers that the name variable is of type string, age is of type int, and salary is of type float64.

We can also use the shorthand syntax := to declare and initialize variables without using the var keyword. For example:

```
name, age, salary := "John", 30, 2500.50
```

2.2. Constants

Golang also supports typed constants. Constants are similar to variables, but their values cannot be changed once they are defined. This means that constants can provide a way to define fixed values that will not be modified during the execution of a program.

Declaring constants

To define a constant in Golang, the **const** keyword is used, followed by the name of the constant and its value. The type of the constant is automatically inferred based on the value that is assigned to it. For example:

```
const pi = 3.14 // declaring a constant named pi with a value of 3.14
```

Once defined, the value of a constant cannot be modified. Attempting to modify the value of a constant will result in a compilation error.

Golang also allows constants to be typed. This means that the type of a constant can be explicitly specified when it is defined. To specify the type of a constant, the type is placed after the constant name, separated by a colon. For example, the following code snippet defines a constant named **myConst** with a type of int:

```
const myConst int = 42
```

Typed constants can be useful when you need to ensure that a constant has a specific type, or when you need to use a constant in a context that requires a specific type.

Declaring multiple constants

We can also declare multiple constants at once like this:

```
const (
    monday = 1
    tuesday = 2
    wednesday = 3
)
```

To declare multiple constants of different data types using a single const clause, you can separate each constant declaration with a comma.

```
const (
    Pi    float64 = 3.14159265358979323846
    Truth bool    = true
```

```
    Name  string  = "Golang"
)
```

In the above example, we have declared three constants of different data types using a single const clause. The first constant Pi is a float64, the second constant Truth is a boolean, and the third constant Name is a string.

Note that when using a single const clause to declare multiple constants, the data type must be specified for each constant.

2.3. Data types

Data types represent the type of data that a variable can hold. The data type of a variable determines the amount of memory space required to store the variable, the range of values that the variable can hold, and the operations that can be performed on the variable.

Golang has a good range of datatypes which can address modern application requirements. Refer to following list of the data types -

Datatype	Size in bytes
bool	1
string	2
int8	1
uint8	1
int16	2
uint16	2
int32	4
uint32	4
int64	8

uint64	8
float32	4
float64	8
complex64	8
complex128	16
byte	1
rune	4

Let's understand each data type in detail along with their applications.

2.3.1 bool

The bool data type is a boolean type that represents a logical value that can only be true or false. In programming, boolean values are commonly used for conditions and branching statements.

Declaring and Assigning bool Variables

To declare a bool variable, we can use the var keyword followed by the variable name and its data type bool. For example:

```
var isTrue bool
```

We can assign a value to a bool variable using the = operator. For example:

```
isTrue = true
```

We can also declare and assign a value to a bool variable in one line using the shorthand declaration :=. For example:

```
isFalse := false
```

Default Value of bool

In Go, the default value of a bool variable is ***false***.

Use Cases

The bool data type is commonly used in programming for conditions and branching statements. For example:

```
if isTrue {
  // do something
} else {
  // do something else
}
```

We can also use bool variables to represent the state of a particular condition or flag.

```
isRunning := true
if isRunning {
  // continue running the program
} else {
  // stop the program
}
```

2.3.2 string

In Golang, the string data type is used to represent a sequence of characters. It is a fundamental data type that is widely used in programming languages. Strings in Golang are represented as a sequence of bytes, with each character in the string represented by a Unicode code point. The string data type is highly versatile and can be used in a variety of applications, from text processing to web development. In this chapter, we will explore the various aspects of the string data type in Golang, including its syntax, functions, and Unicode support.

Declaring and Assigning string Variables

In Golang, a string is a sequence of Unicode characters. To create a string, you can use double quotes ("") or backticks (``) to enclose the characters.

```
str1 := "Hello, World!"

str2 := `Hello,

      World!`
```

In the example above, we created two strings, str1 and str2. str1 is created using double quotes and str2 is created using backticks. Backticks are often used for multiline strings or for strings that contain special characters, such as backslashes or quotes.

Default Value of string

The default value of a string data type is an empty string "". This means that if a string variable is declared without an initial value, it will automatically be assigned an empty string as its default value. It is important to note that an empty string is different from a nil value, which is the default value for most other data types in Golang. An empty string represents a zero-length string, whereas nil represents the absence of a value.

Working with String Methods and Functions

Golang provides many built-in functions and methods for working with strings. Here are some commonly used string **functions**:

- **len**(str) returns the length of the string in bytes.
- strings.**Contains**(str, substr) returns true if the string str contains the substring substr.

- strings.**Index**(str, substr) returns the index of the first occurrence of the substring substr in the string str.
- strings.**Replace**(str, old, new, n) replaces n occurrences of the substring old with the substring new in the string str.

Here are some commonly used string **methods**:

- str**[i]** returns the character at index i in the string str.
- str**[start:end]** returns the substring of str starting from index start up to (but not including) index end.
- strings.**ToLower**() and strings.**ToUpper**() return a new string with all characters converted to lowercase or uppercase, respectively.
- strings.**TrimSpace**() returns a new string with all leading and trailing white space removed.

Concatenating Strings

In Golang, you can concatenate strings using the + operator or the fmt.Sprintf() function. Here's an example:

```
str1 := "Hello"
str2 := "World"
str3 := str1 + " " + str2 // Concatenate strings using the + operator
str4 := fmt.Sprintf("%s %s", str1, str2) // Concatenate strings using fmt.Sprintf()
```

Comparing Strings

In Golang, you can compare strings using the == and != operators. Here's an example:

```
str1 := "Hello"
str2 := "World"
if str1 == str2 {
    fmt.Println("The strings are equal.")
} else {
    fmt.Println("The strings are not equal.")
}
```

Use cases

The string data type is one of the most commonly used data types in programming, and for good reason. It can be used to represent a wide variety of data, including text, numbers, and even binary data. Some of the most common use cases for the string data type include:

Storing user input: Whenever your program needs to receive input from a user, it will likely be in the form of a string. Whether it's a name, an email address, or a message, the string data type is the most natural way to store this kind of data.

Manipulating text: If your program needs to work with text in any way, you will need to use the string data type. This can include tasks such as searching for specific characters or substrings, replacing parts of a string, or formatting text for display.

File I/O: When working with files, you'll often need to read or write strings. For example, if you're writing a program that reads a text file and performs some analysis on its contents, you'll likely use the string data type to represent the file's contents.

Network communication: When sending or receiving data over a network, the string data type is often used. This could include sending messages over a chat application, transmitting data between servers, or even just displaying data on a web page.

In short, the string data type is an essential tool for any programmer working with text or data that can be represented as text. Its versatility and ease of use make it a key component of many programming languages, including Golang.

Unicode Support

Golang has excellent support for Unicode, which is reflected in its string handling capabilities. In Golang, strings are made up of a sequence of Unicode code points, which are 32-bit unsigned integers that represent each character in the string. This means that Golang can handle not just ASCII characters, but also characters from other languages, such as Chinese, Japanese, and Arabic.

One important thing to note is that in Golang, a string is actually an immutable sequence of bytes. However, Golang's string type is designed to handle Unicode code points transparently, making it easy to work with Unicode strings in Golang.

To create a string literal in Golang, you simply enclose the text in double quotes. For example:

```
str := "Hello, 世界"
```

This creates a string that contains the text "Hello, 世界", which includes both ASCII characters and Chinese characters.

Golang provides a rich set of functions and methods for working with strings, such as concatenation, splitting, trimming, and more. These functions are designed to handle Unicode strings correctly, ensuring that your application can work with text in any language.

Overall, Golang's strong support for Unicode makes it a great choice for applications that need to work with text in multiple languages. Whether you're building a web application that needs to handle user input in different languages, or a desktop application that needs to support internationalization, Golang's string handling capabilities have got you covered.

2.3.3 int

One of the most commonly used data types in Golang is the integer data type, abbreviated as "int".

An integer is a whole number that can be either positive or negative. The size of an integer depends on the platform where the code is running. In Golang, the size of an integer is 32 bits on a 32-bit platform and 64 bits on a 64-bit platform.

Golang supports both signed(int) and unsigned(uint) integers. Signed integers are used to represent negative and positive values, while unsigned integers are used to represent only positive values. These subtypes include:

- int8: An 8-bit signed integer, with a minimum value of -128 and a maximum value of 127.
- int16: A 16-bit signed integer, with a minimum value of -32768 and a maximum value of 32767.
- int32: A 32-bit signed integer, with a minimum value of -2147483648 and a maximum value of 2147483647.
- int64: A 64-bit signed integer, with a minimum value of -9223372036854775808 and a maximum value of 9223372036854775807.

In addition, Golang also has unsigned integer types that use the same number of bits as their signed counterparts, but only store non-negative values. These subtypes include:

- uint8 (or byte): An 8-bit unsigned integer, with a minimum value of 0 and a maximum value of 255.
- uint16: A 16-bit unsigned integer, with a minimum value of 0 and a maximum value of 65535.
- uint32: A 32-bit unsigned integer, with a minimum value of 0 and a maximum value of 4294967295.
- uint64: A 64-bit unsigned integer, with a minimum value of 0 and a maximum value of 18446744073709551615.

These subtypes can be used to optimize memory usage, depending on the range of values expected for a particular variable. For example, if a variable will only ever store values between 0 and 255, it would be more efficient to use an uint8 instead of an int32. However, it's important to note that operations between different subtypes can result in unexpected behavior and should be handled with care.

Declaring and Assigning string Variables

To declare an int variable, we use the "var" keyword followed by the variable name and its initial value. Here is an example:

```
var a int = 10
```

This declares an integer variable named "a" and initializes it with the value of 10. If we don't specify the initial value, Golang will set it to the default value of 0.

We can also declare multiple integer variables using the same var keyword, separating each variable with a comma. Here is an example:

```
var x, y int = 10, 20
```

This declares two integer variables named "x" and "y" and initializes them with the values of 10 and 20, respectively.

Default Value of int

The default value of an int type in Go is 0. When a variable of int type is declared without assigning a value, it is automatically initialized to 0, as Go initializes all numeric variables to 0 by default.

Use cases

The int data type is commonly used to store integer values in programming. It is a versatile data type that can be used in a wide range of applications. For example -

- It can be used to store the number of items in a shopping cart or the number of users on a website.
- The int data type is also useful for performing mathematical operations such as addition, subtraction, multiplication, and division.

- In addition, it can be used for creating loops and iterating through arrays or other data structures, also be used as indices in arrays or slices to access specific elements.

Overall, the int data type is an essential building block of any Go program and is used extensively in various applications.

2.3.4 float

A float data type is a variable type that represents a single-precision floating-point number. The float data type in Go is represented by a 32-bit binary value, which allows it to store a wide range of values.

There are two subtypes of the float data type: float32 and float64, both of which are signed data types. An unsigned float data type does not exist in Go.

float32 is a 32-bit floating-point number, which means it can represent a wide range of values, from approximately -3.4 x 10^38 to +3.4 x 10^38. However, because it has fewer bits than float64, it has less precision, meaning that it can represent fewer decimal places. In general, float32 is suitable for most applications that require high-precision calculations, such as scientific or financial calculations, but not for cases where extreme precision is needed.

float64, on the other hand, is a 64-bit floating-point number, which means it has a larger range and higher precision than float32. float64 can represent values from approximately -1.8 x 10^308 to +1.8 x 10^308, and can store more decimal places than float32. This makes it suitable for applications that require extremely high precision, such as astronomy, physics, and other scientific fields where calculations require a high degree of accuracy.

Declaring and Assigning float Variables

You can declare and assign float variables using the "float32" data type. Here's an example of how to declare and assign a float variable in Go:

```
var myFloat float32

myFloat = 3.14
```

In the above code, we first declare a variable called "myFloat" of type "float32". We then assign the value of 3.14 to this variable using the assignment operator "=".

Alternatively, you can declare and assign a float variable in Go using the shorthand syntax:

```
myFloat := 3.14
```

In the above code, we use the ":=" operator to declare and assign the value of 3.14 to the variable "myFloat". This shorthand syntax is often used in Go for concise and readable code.

You can also declare and assign multiple float variables in a single line of code:

```
var x, y, z float32 = 1.0, 2.0, 3.0
```

In the above code, we declare three float variables "x", "y", and "z" of type "float32" and assign them the values of 1.0, 2.0, and 3.0, respectively.

It is important to note that Go is a statically-typed language, which means that the data type of a variable must be declared at the time of variable creation. This ensures that the compiler can catch any type mismatches or errors during compile-time rather than run-time.

Default Value of float

The default value of a float variable is 0. This means that if you declare a float variable but do not initialize it with a value, it will be automatically assigned the value of 0. For example:

```
var myFloat float64

fmt.Println(myFloat) // Output: 0
```

In the above code, we declare a float variable "myFloat" of type "float64" but do not assign it a value. When we print the value of "myFloat" to the console using the fmt.Println() function, the output will be 0, which is the default value for a float variable in Go.

It is important to note that while the default value of a float variable is 0, it is generally good practice to always initialize variables with a value before using them in your program. This helps to prevent unexpected behavior or errors that may occur due to uninitialized variables.

Use cases

The float data type in Go programming language is commonly used in a variety of applications that require the representation of decimal values with a fractional component. Here are some common use cases for the float data type in Go:

- Financial applications: Float data types are commonly used in financial applications to represent currency values, interest rates, and other financial metrics that require high precision.
- Scientific and engineering applications: Float data types are used in scientific and engineering applications to represent physical measurements, such as temperature, weight, distance, and time.
- Gaming and animation: Float data types are used in gaming and animation applications to represent the position, rotation, and movement of objects in a 3D space.
- Statistical analysis: Float data types are used in statistical analysis to represent numerical data, such as averages, standard deviations, and other statistical measures.
- Machine learning: Float data types are used in machine learning algorithms to represent the weights and biases of neural networks, which are used to make predictions based on input data.

In general, the float data type is useful in any application that requires the representation of decimal values with a fractional component. It is important to choose the appropriate data type based on the range and precision required for your specific use case. If high precision is required, float64 is typically the best choice, while float32 may be sufficient for applications where memory usage is a concern.

2.3.5 complex

The complex data type is a built-in data type that represents complex numbers in rectangular form. A complex number consists of two components: a real part and an imaginary part. The real part is a float64 value, and the imaginary part is also a float64 value, but it is suffixed with the "i" character to denote that it is an imaginary number.

The complex data type in Go is defined as follows:

```
type complex64 complex128

type complex128 complex128
```

The complex data type is implemented using two float64 values, one for the real part and one for the imaginary part. The complex64 data type uses float32 values for the real and imaginary parts, while the complex128 data type uses float64 values for both parts.

Declaring and Assigning complex Variables

To declare and initialize a complex variable in Go, we use the "complex" function, which takes two float64 arguments representing the real and imaginary parts respectively. Here's an example:

```
var myComplex complex128 = complex(3.0, 4.0)
```

In the above code, we declare a complex variable "myComplex" of type "complex128" and initialize it with the "complex" function, which sets the real part to 3.0 and the imaginary part to 4.0.

Go provides several built-in functions to work with complex numbers, including "real" and "imag", which return the real and imaginary parts of a complex number, respectively. Here's an example:

```
var myComplex complex128 = complex(3.0, 4.0)
fmt.Println(real(myComplex)) // Output: 3
fmt.Println(imag(myComplex)) // Output: 4
```

In the above code, we declare a complex variable "myComplex" and use the "real" and "imag" functions to print out its real and imaginary parts to the console.

Default Value of complex

The default value of a complex variable is 0+0i. This means that if you declare a complex variable but do not initialize it with a value, it will be automatically assigned the value of 0+0i. For example:

```
var myComplex complex64
fmt.Println(myComplex) // Output: (0+0i)
```

In the above code, we declare a complex variable "myComplex" of type "complex64" but do not assign it a value. When we print the value of "myComplex" to the console using the fmt.Println() function, the output will be (0+0i), which is the default value for a complex variable in Go.

Use Cases

The complex data type in Go programming language is commonly used in applications that require the representation and manipulation of complex numbers. Here are some common use cases for the complex data type in Go:

- Engineering and scientific applications: The complex data type is widely used in engineering and scientific applications to represent complex numbers in fields such as control systems, signal processing, electromagnetics, and quantum mechanics.

- Mathematics: The complex data type is used extensively in mathematical applications that involve complex numbers, including number theory, algebra, and geometry.
- Computer graphics and animation: The complex data type is used in computer graphics and animation to represent the position, rotation, and movement of objects in a 3D space.
- Cryptography: The complex data type is used in some cryptographic algorithms that rely on complex number arithmetic, such as elliptic curve cryptography.
- Machine learning: The complex data type is used in some machine learning algorithms that involve complex number arithmetic, such as quantum machine learning.

In general, the complex data type is useful in any application that requires the representation and manipulation of complex numbers. It is important to choose the appropriate data type based on the range and precision required for your specific use case. If high precision is required, complex128 is typically the best choice, while complex64 may be sufficient for applications where memory usage is a concern.

2.3.6 byte

The byte data type is a built-in data type that represents a single byte of data. It is an alias for the uint8 data type, which means that it can store integer values between 0 and 255.

Declaring and Assigning byte Variables

To declare a byte variable in Go, you can use the following syntax:

```
var b byte
```

In the above code, we declare a byte variable "b" without initializing it. By default, the value of "b" will be 0.

You can also initialize a byte variable with a value using the following syntax:

```
var b byte = 255
```

In the above code, we declare a byte variable "b" and initialize it with the value of 255.

Default Value of byte

In Go, a byte is an alias for the uint8 type, which is an unsigned 8-bit integer. By default, when a byte variable is declared, its initial value is set to 0. For example, consider the following code snippet:

```
var b byte
fmt.Println(b)
```

This will output 0 to the console, indicating that the variable b has been initialized to its default value of 0.

Similarly, if a byte variable is declared within a struct or an array, all of its elements will also be initialized to 0 by default.

Use Cases

In Go, the byte data type is an alias for the uint8 type, which represents an unsigned 8-bit integer. This data type has several use cases in Go programming, including:

- Encoding and Decoding: Byte data type is widely used in encoding and decoding data. Many encoding and decoding algorithms, such as base64 and hex encoding, operate on byte data. Therefore, byte data type is essential in implementing these algorithms.
- Network Programming: When sending data over the network, the data is usually represented in bytes. Thus, byte data type is widely used in network programming to send and receive data.

- File Operations: Byte data type is also used in file operations, such as reading and writing binary files. In file operations, the data is usually represented in bytes, and thus byte data type is essential in handling such operations.
- Image Processing: In image processing applications, the image data is often represented in bytes. Byte data type is widely used in handling image data and in implementing various image processing algorithms.
- Cryptography: Byte data type is also essential in implementing various cryptography algorithms. Many cryptographic algorithms, such as AES and SHA, operate on byte data. Therefore, byte data type is crucial in implementing these algorithms.
- Buffering: Byte data type is widely used in buffering data. In many cases, data is read from a source and temporarily stored in a buffer, and the buffer is then used to process the data. Byte data type is used in implementing such buffers.

In conclusion, the byte data type has several use cases in Go programming, including encoding and decoding, network programming, file operations, image processing, cryptography, and buffering. The versatility of the byte data type makes it an essential tool for handling a wide range of data in Go programming.

2.3.7 rune

The rune data type represents a Unicode code point. It is an alias for the int32 type and can store values from 0 to 0x10FFFF, which is the maximum value of a Unicode code point.

The rune data type is primarily used for working with Unicode strings, as it provides a convenient way to represent individual characters in a string. For example, a string in Go can be represented as a slice of rune values, allowing individual characters to be easily accessed and manipulated.

One of the main benefits of using the rune data type in Go is that it provides full support for Unicode characters. This means that any Unicode character can be represented and manipulated using the rune data type. This is important when working with internationalization and localization in software development, where it is necessary to support multiple languages and character sets.

It's worth noting that the rune data type can also be used as an alias for int32 in contexts where the value represents a character code, but it is generally recommended to use rune specifically for working with Unicode code points.

Declaring and Assigning rune Variables

To declare a rune variable, you can use the var keyword followed by the variable name and the rune type, like this:

```
var r rune
```

This declares a rune variable named r with an initial value of 0. You can also assign an initial value to the variable when you declare it, like this:

```
var r rune = 'a'
```

This declares a rune variable named r with an initial value of the Unicode code point for the lowercase letter a, which is 97.

You can assign a new value to a rune variable using a single-quoted character literal or a Unicode escape sequence, like this:

```
r = 'b'
```

This assigns the Unicode code point for the lowercase letter b, which is 98, to the r variable. Alternatively, you can use a Unicode escape sequence like this:

```
r = '\u00e9'
```

This assigns the Unicode code point for the lowercase letter é, which is U+00E9, to the r variable.

Default Value of rune

The default value of a rune variable is 0. When a rune variable is declared without an initial value, it is automatically initialized to 0. For example, the following code declares a rune variable r without an initial value:

```
var r rune
fmt.Println(r) // Output: 0
```

Use Cases

Here are some common use cases for the rune data type in Golang:

- Parsing and processing text: When working with text data that contains non-ASCII characters, the rune data type can be used to represent individual characters and perform various text processing operations.
- Regular expressions: Regular expressions are a powerful tool for pattern matching and text processing. Since Go supports Unicode characters and strings, the rune data type can be used to match specific Unicode characters or character ranges in regular expressions.
- Database applications: When working with databases that store text data in different character sets or encodings, the rune data type can be used to represent Unicode characters in a consistent manner.
- File I/O operations: When reading or writing text data from files, the rune data type can be used to represent individual characters and perform various text processing operations.

In summary, the rune data type is an important data type in Golang that is used to represent Unicode code points. It is useful in a variety of applications that involve text processing, regular expressions, database operations, and file I/O.

2.4. Operators

Operators in programming are special symbols or keywords that execute specific operations on one or more pieces of data. Golang features a variety of operators, similar to other programming languages, that allow you to manipulate data effectively. Some of the most frequently used operators in Golang include:

- **Arithmetic Operators:** These are used to perform basic mathematical operations such as addition, subtraction, multiplication, and division. The symbols used for these operations are +, -, *, and / respectively.
- **Relational Operators:** These are used to compare two values and check their relationship with each other. They are also called comparison operators. The symbols used for these operations are <, >, <=, >=, ==, and !=. For example, x > y checks if the value of x is greater than the value of y.
- **Logical Operators:** These are used to combine or negate the logical values of two or more expressions. The symbols used for these operations are && (logical AND), || (logical OR), and ! (logical NOT).
- **Assignment Operators:** These are used to assign a value to a variable. The symbol used for this operation is =. For example, x = 10 assigns the value of 10 to the variable x.
- **Bitwise Operators:** These are used to perform operations on individual bits of a number. The symbols used for these operations are &, |, ^, <<, and >>. For example, x & y performs a bitwise AND operation on the values of x and y.
- **Other Operators:** Golang also includes several other operators, such as the increment (++) and decrement (--) operators used to increase or decrease the value of a variable by 1, and the ternary operator (?:) used for conditional statements.

These operators are essential building blocks for any Golang program, and their proper use can significantly improve the efficiency and functionality of the program.

2.4.1 Arithmetic Operators

The arithmetic operators are the foundation of mathematical operations on numeric data in Golang.

1. **Addition Operator (+)**: The addition operator (+) is used to add two values together. It can be used with all numeric data types, including int, float, and complex numbers. Here's an example:

```
a := 10
b := 5
c := a + b // c = 15
```

2. **Subtraction Operator (-):** The subtraction operator (-) is used to subtract one value from another. It can be used with all numeric data types, including int, float, and complex numbers. Here's an example:

```
a := 10
b := 5
c := a - b // c = 5
```

3. **Multiplication Operator (*):** The multiplication operator () is used to multiply two values together. It can be used with all numeric data types, including int, float, and complex numbers. Here's an example:

```
a := 10
b := 5
c := a * b // c = 50
```

4. **Division Operator (/):** The division operator (/) is used to divide one value by another. It can be used with all numeric data types, including int, float, and complex numbers. Here's an example:

```
a := 10
b := 5
```

```
c := a / b // c = 2
```

Note that if you divide an integer by an integer, the result will be an integer. If you want a float result, you should use float values in the division operation.

5. **Modulus Operator (%):** The modulus operator (%) is used to find the remainder of a division operation. It can be used with integer data types only. Here's an example:

```
a := 10
b := 3
c := a % b // c = 1
```

In this example, the modulus operator is used to find the remainder when 10 is divided by 3.

Order of operation

The arithmetic operators follow the standard order of operations, which means that multiplication and division are performed before addition and subtraction. However, we can use parentheses to override this order and specify which operation should be performed first.

```
package main
import "fmt"
func main() {
 // using parentheses to override order of operations
 var x int = 10
 var y int = 5
 var z int = (x + y) * 2
```

```
  fmt.Println(z) // Output: 30
}
```

In the above example, we have used parentheses to specify that addition should be performed before multiplication.

It is also important to note that division by zero will result in a runtime error. Therefore, we should always check for a zero divisor before performing division.

```
package main
import "fmt"
func main() {

  // checking for zero divisor
  var x int = 10
  var y int = 0

  if y == 0 {
    fmt.Println("Cannot divide by zero!")
  } else {
    var z int = x / y
    fmt.Println(z)
  }
}
```

In the above example, we have used an if statement to check if the divisor is zero before performing division. If the divisor is zero, we print an error message. Otherwise, we perform division and print the result.

2.4.2 Relational Operators

The Relational operators in Go are used to compare two values and return a boolean value (true or false) based on the comparison result.

1. **Equal to (==):** The *equal to* operator checks if two values are equal or not. It returns true if the values are equal, and false otherwise.

```
x := 5
y := 5
if x == y {
  fmt.Println("x is equal to y")
}
```

2. **Not equal to (!=):** The *not equal to* operator checks if two values are not equal. It returns true if the values are not equal, and false if they are equal.

```
x := 5
y := 10
if x != y {
  fmt.Println("x is not equal to y")
}
```

3. **Greater than (>):** The *greater than* operator checks if the first value is greater than the second value. It returns true if the first value is greater, and false otherwise.

```
x := 10
y := 5
if x > y {
```

```
    fmt.Println("x is greater than y")
}
```

4. **Less than (<):** The *less than* operator checks if the first value is less than the second value. It returns true if the first value is less, and false otherwise.

```
x := 5
y := 10
if x < y {
    fmt.Println("x is less than y")
}
```

5. **Greater than or equal to (>=):** The *greater than or equal to* operator checks if the first value is greater than or equal to the second value. It returns true if the first value is greater or equal, and false otherwise.

```
x := 10
y := 5
if x >= y {
    fmt.Println("x is greater than or equal to y")
}
```

6. **Less than or equal to (<=):** The *less than or equal to* operator checks if the first value is less than or equal to the second value. It returns true if the first value is less or equal, and false otherwise.

```
x := 5
```

```
y := 10
if x <= y {
    fmt.Println("x is less than or equal to y")
}
```

2.4.3 Logical Operators

The logical operators are used to perform logical operations on boolean values. The three logical operators are AND (&&), OR (||), and NOT (!).

1. **AND Operator (&&):** The AND operator (&&) returns true if both its operands evaluate to true, and false otherwise. The following truth table shows the possible combinations of operand values and their results:

Operand 1	Operand 2	Result
true	true	true
true	false	false
false	true	false
false	false	false

Here's an example that uses the AND operator:

```
x := 5
y := 10
if x > 0 && y > 0 {
    fmt.Println("Both x and y are positive")
}
```

In this example, the && operator is used to test whether both x and y are positive. If both are positive, the message "Both x and y are positive" is printed to the console.

2. **OR Operator (||):** The OR operator (||) returns true if at least one of its operands evaluates to true, and false otherwise. The following truth table shows the possible combinations of operand values and their results:

Operand 1	Operand 2	Result
true	true	true
true	false	true
false	true	true
false	false	false

Here's an example that uses the OR operator:

```
x := 5
y := -10
if x > 0 || y > 0 {
    fmt.Println("At least one of x and y is positive")
}
```

In this example, the || operator is used to test whether at least one of x and y is positive. Since x is positive, the message "At least one of x and y is positive" is printed to the console.

3. **NOT Operator (!):** The NOT operator (!) returns the opposite of its operand. If the operand is true, ! returns false, and if the operand is false, ! returns true.

Here's an example that uses the NOT operator:

```
x := 5
if !(x > 10) {
    fmt.Println("x is less than or equal to 10")
}
```

In this example, the ! operator is used to test whether x is not greater than 10. Since x is less than or equal to 10, the message "x is less than or equal to 10" is printed to the console.

2.4.4 Assignment Operators

The assignment operators are used to assign values to variables. The basic assignment operator is the equal sign (=). There are also compound assignment operators that combine arithmetic or bitwise operations with the assignment operator. These operators provide a more concise way to perform common operations. The following are the compound assignment operators in Go:

1. **Addition assignment (+=):** adds the value of the right-hand operand to the value of the left-hand operand and assigns the result to the left-hand operand.

```
var x int = 10
x += 5 // equivalent to x = x + 5, assigns the value 15 to x
```

2. **Subtraction assignment (-=):** subtracts the value of the right-hand operand from the value of the left-hand operand and assigns the result to the left-hand operand.

```
var x int = 10
x -= 5 // equivalent to x = x - 5, assigns the value 5 to x
```

3. **Multiplication assignment (*=):** multiplies the value of the right-hand operand by the value of the left-hand operand and assigns the result to the left-hand operand.

```
var x int = 10

x *= 5 // equivalent to x = x * 5, assigns the value 50 to x
```

4. **Division assignment (/=):** divides the value of the left-hand operand by the value of the right-hand operand and assigns the result to the left-hand operand.

```
var x int = 10

x/= 5 // equivalent to x = x/ 5, assigns the value 2 to x
```

5. **Modulo assignment (%=):** calculates the remainder of the division of the left-hand operand by the right-hand operand and assigns the result to the left-hand operand.

```
var x int = 10

x %= 3 // equivalent to x = x % 3, assigns the value 1 to x
```

6. **Bitwise AND assignment (&=):** performs a bitwise AND operation between the left-hand operand and the right-hand operand and assigns the result to the left-hand operand.

```
var x uint8 = 0x0F

x &= 0x03 // equivalent to x = x & 0x03, assigns the value 0x03 to x
```

7. **Bitwise OR assignment (|=):** performs a bitwise OR operation between the left-hand operand and the right-hand operand and assigns the result to the left-hand operand.

```
var x uint8 = 0x0F
x |= 0x30 // equivalent to x = x | 0x30, assigns the value 0x3F to x
```

8. **Bitwise XOR assignment (^=):** performs a bitwise XOR operation between the left-hand operand and the right-hand operand and assigns the result to the left-hand operand.

```
var x uint8 = 0x0F
x ^= 0x33 // equivalent to x = x ^ 0x33, assigns the value 0x3C to x
```

9. **Bitwise left shift assignment (<<=):** performs a bitwise left shift operation on the left-hand operand by the number of bits specified by the right-hand operand and assigns the result to the left-hand operand.

```
var x uint8 = 2
x <<= 2 // equivalent to x = x << 2, assigns the value 8 to x
```

10. **Bitwise right shift assignment (>>=):** performs a bitwise right shift operation on the left-hand operand by the number of bits specified by the right-hand operand and assigns the result to the left-hand operand.

```
var x uint8 = 8
x >>= 2 // equivalent to x = x >> 2, assigns the value 2 to x
```

2.4.5 Bitwise Operators

The bitwise operators are used to perform operations on individual bits of integer values.

1. **Bitwise AND (&):** Performs a bitwise AND operation on the two operands, resulting in a new value where each bit is set to 1 only if both corresponding bits in the operands are also 1.

```
a := 5      // binary representation: 0000 0101
b := 3      // binary representation: 0000 0011
result := a & b // binary representation: 0000 0001
fmt.Println(result) // Output: 1
```

2. **Bitwise OR (|):** Performs a bitwise OR operation on the two operands, resulting in a new value where each bit is set to 1 if either or both corresponding bits in the operands are also 1.

```
a := 5      // binary representation: 0000 0101
b := 3      // binary representation: 0000 0011
result := a | b // binary representation: 0000 0111
fmt.Println(result) // Output: 7
```

3. **Bitwise XOR (^):** Performs a bitwise XOR (exclusive OR) operation on the two operands, resulting in a new value where each bit is set to 1 only if exactly one of the corresponding bits in the operands is 1.

```
a := 5      // binary representation: 0000 0101
b := 3      // binary representation: 0000 0011
result := a ^ b // binary representation: 0000 0110
fmt.Println(result) // Output: 6
```

4. **Bitwise NOT (~):** Performs a bitwise NOT operation on the operand, resulting in a new value where each bit is inverted (i.e., 0s become 1s and 1s become 0s).

```
var a uint8 = 42 // 00101010 in binary
var b uint8 = ^a // 11010101 in binary (bitwise complement of a)
fmt.Printf("a: %08b\n", a) // prints "a: 00101010"
fmt.Printf("b: %08b\n", b) // prints "b: 11010101"
```

5. **Left shift (<<):** Shifts the bits of the left-hand operand to the left by the number of positions specified by the right-hand operand. The vacated bits are set to 0.

```
a := 5      // binary representation: 0000 0101
result := a << 2// binary representation: 0001 0100
fmt.Println(result) // Output: 20
```

6. **Right shift (>>):** Shifts the bits of the left-hand operand to the right by the number of positions specified by the right-hand operand. For unsigned integers, the vacated bits are set to 0, while for signed integers, the value of the most significant bit (the sign bit) is preserved.

```
a := 5      // binary representation: 0000 0101
result := a >> 2// binary representation: 0000 0001
fmt.Println(result) // Output: 1
```

2.5. Functions

Functions are a cornerstone of Go programming, facilitating code reuse and modularization. They allow you to encapsulate specific functionality within named blocks of code, which can then be invoked elsewhere in your application.

2.5.1 Definition of Functions

In Go, a function is defined with the `` `func` `` keyword, followed by the function's name, a pair of parentheses that may contain parameters, and a set of curly braces **{}** that encloses the function's body. Here's the basic syntax:

```
func functionName(parameter1 type1, parameter2 type2)
returntype {
    // function body
}
```

Functions can accept zero or more parameters and can return one or more values.

2.5.2 Calling Functions

Functions are called by their name followed by parenthesis containing any required arguments. For instance, a function **sum** accepting two integers would be called as **sum(5, 3)**.

```
// function definition
func sum() {
   fmt.Println("Function sum called")
}
func main() {
   // function call
   sum(5, 3)
}
```

2.5.3 Returning Functions

Go allows functions to return multiple values, which can be particularly useful for error handling. To specify multiple return types, encapsulate them in parentheses. For example:

```
func sum(a, b int) (int) {
   return a + b
}
```

Named Return Values:

Go also supports named return values. This means you can specify the names of the return values when you define your function, and then simply use these names in your function body. The return statement without explicit values returns the named return values. This is known as a "naked" return. Here's an example:

```
func calculate(a int, b int) (sum int) {
   sum = a + b
   return
}
```

In this example, calculate is a function with named return values sum. In the function body, we compute the sum of the input arguments and assign the results to sum. The return statement will automatically return the value stored in the sum.

Order of Return Values:

The order of return values in Go functions is important and should be consistent with how they are declared in the function signature. When multiple values are returned, they are returned in the order that they're defined in the function signature.

```
func multipleReturns(a int) (int, string) {
    return a * 2, "multiplied by 2"
}
func main() {
    val, str := multipleReturns(5)
    fmt.Println(val, str)  // prints: 10 multiplied by 2
}
```

In this example, the function **'multipleReturns'** is declared with two return types: **'int'** and **'string'**. When calling this function, the first returned value is expected to be an **'int'** and the second a **'string'**.

So when receiving multiple return values from a function in Go, you should assign them to variables in the same order as the function's return signature. This consistency in ordering helps avoid bugs and makes your code easier to read and understand.

2.5.4 Anonymous Functions

Anonymous functions, often known as lambda functions in other languages, are a fundamental aspect of function literals in Go. They are called anonymous because they aren't declared with a standard function definition (they don't have a name). Instead, anonymous functions are inline functions declared with a function literal.

Go supports anonymous functions, which can form closures. Anonymous functions are useful when you want to define a function inline without having to name it.

```
package main
import "fmt"
func main() {
```

```
    sum := func(a int, b int) int {
        return a + b
    }
    fmt.Println(sum(3, 4)) // prints: 7
}
```

In this example, an anonymous function is created that adds two integers together. The function is assigned to the variable sum, and can then be called using **sum**(3, 4).

Anonymous functions are very useful when you want to create an inline function that won't be used elsewhere. They can also be used as arguments to other functions, or to immediately self-execute a block of code. In Go, they are often used with Go's concurrency primitives or when creating small chunks of transformative code.

IIFE

Go supports the use of immediately invoked function expressions (IIFE), an anonymous function that is executed as soon as it is defined. Example:

```
package main
import "fmt"
func main() {
    func(msg string) {
        fmt.Println(msg)
    }("Hello, World!") // prints: Hello, World!
}
```

In this example, the anonymous function is immediately invoked with the string "Hello, World!" as an argument.

Chapter 3
Control Structures in Go

Control structures are fundamental elements in programming that guide the flow of execution of a program. They are the building blocks that enable a programmer to dictate how different pieces of code are executed based on specified conditions, and they play a crucial role in creating dynamic and functional programs.

Control structures are constructs used in programming languages to control the flow and order of execution of statements in a script or program. They determine the direction the program takes based on certain conditions or the need to repeatedly execute a block of code.

Imagine you're following a recipe to bake a cake, which is a series of steps that must be followed to achieve the desired outcome. However, not all steps are straightforward. For instance:

- **Checking if ingredients are available:** Before you start baking, you need to check if you have all the necessary ingredients like flour, eggs, and sugar. If something is missing, you might decide to buy it from the store or change the recipe.
- **Mixing ingredients:** If all ingredients are present, you proceed to mix them. But how long you mix them could depend on the type of cake you're making.

- **Baking:** Once the batter is ready, you bake it. The baking time and temperature can vary depending on the cake type. For example, a small cupcake will bake faster than a thick pound cake.
- **Checking if the cake is ready:** You don't just wait for a timer; you might check if the cake is done by inserting a toothpick in the center. If it comes out clean, the cake is ready. If not, you continue baking and check again later.

Importance of Control Structures

Decision Making: Control structures allow programs to make decisions and execute different code based on varying conditions. This decision-making capability is essential for handling real-world scenarios in programming.

Code Reusability and Reduction: They help in avoiding repetition of code by handling tasks that need to be repeated. This not only saves time but also makes the code more readable and maintainable.

Program Flexibility: With control structures, programs can be written to adapt to different inputs or situations, making the program more flexible and dynamic.

Complexity Management: Control structures aid in managing and reducing the complexity of programs. They allow the breaking down of complicated tasks into simpler, manageable sub-tasks.

Understanding these control structures is essential for any programmer working with Go, as they form the backbone of most programming tasks. The upcoming sections will delve into each of these structures in detail, providing a deeper understanding of how they can be effectively used in Go programming.

3.1 Conditional Statements

Conditional statements are a critical feature in programming, allowing for decision-making processes within a code. They enable a program to choose different paths of execution based on specific conditions, making the program dynamic and responsive to different inputs or situations. In essence, conditional statements evaluate a condition or expression, and based on whether it's true or false, they direct the program to execute certain blocks of code accordingly.

In Go, conditional statements are a cornerstone for decision-making in programs. They assess conditions and execute specific blocks of code accordingly. The if, else if, and else statements are used extensively. For example, an if statement can validate user input. This section delves into the syntax and practical applications, emphasizing their critical role in programming logic.

3.1.1 If Statement

The if statement in Go is a fundamental control structure used for decision-making. It allows you to execute a block of code conditionally based on whether a specified condition evaluates to true. The simplicity and versatility of the if statement make it one of the most commonly used control structures in Go programming.

Syntax

The basic syntax of an if statement in Go is as follows:

```
if condition {
    // code to execute if condition is true
}
```

Here, condition is an expression that evaluates to a boolean value (true or false). If the condition is true, the code inside the curly braces {} is executed. If the condition is false, the code block is skipped.

Features of the If Statement in Go

- *No Parentheses Around Conditions:* Unlike some other languages, Go does not require parentheses around the condition in an if statement.
- *Initialisation Statement*: Optionally, you can include an initialization statement before the condition. This statement often declares and initializes a variable used in the condition.

Example

Let's look at an example to illustrate the use of an if statement in Go:

```
package main
import "fmt"
func main() {
   number := 10
   if number > 5 {
      fmt.Println("Number is greater than 5")
   }
}
// Output: Number is greater than 5
```

In this example, we have a variable number with a value of 10. The if statement checks if the number is greater than 5. Since this condition is true, the program prints "Number is greater than 5".

Using Initialization Statement

Here's an example with an initialization statement:

```
package main
import "fmt"
func main() {
   if number := 10; number > 5 {
      fmt.Println("Number is greater than 5")
   }
```

```
}
// Output: Number is greater than 5
```

In this example, number := 10 is the initialization statement. It declares and initializes a number, which is then used in the conditional check.

3.1.2 If-Else Statement

The if-else statement in Go is a control structure that extends the functionality of the if statement by providing an alternative path of execution when the if condition is not met. This allows a programmer to define two distinct blocks of code: one that executes if a condition is true, and another that executes if the condition is false.

Syntax

The basic syntax of an if-else statement in Go is as follows:

```
if condition {
    // code to execute if condition is true
} else {
    // code to execute if condition is false
}
```

In this structure, condition is a boolean expression. If condition evaluates to true, the block of code inside the first set of curly braces {} is executed. If condition evaluates to false, the code inside the else block is executed.

Features of the If-Else Statement in Go

- *Binary Decision Making:* The if-else statement in Go is used for binary decision-making, where there are only two possible paths of execution.
- *No Parentheses Required:* Like the if statement, Go does not require parentheses around the condition.

- *Initialization Statement:* The if-else statement in Go can also include an initialization statement as part of the if condition.

Example

Let's look at an example to understand the if-else statement in Go:

```
package main
import "fmt"
func main() {
   number := 8
   if number%2 == 0 {
      fmt.Println("Number is even")
   } else {
      fmt.Println("Number is odd")
   }
}
// Output: Number is even
```

In this example, the program checks whether the variable number is even or odd. The if condition number%2 == 0 checks if number is divisible by 2. If it is, the program prints "Number is even"; otherwise, it prints "Number is odd".

Using Initialization Statement

Here's an example with an initialization statement:

```
package main
import "fmt"
func main() {
    if number := 7; number%2 == 0 {
        fmt.Println("Number is even")
    } else {
        fmt.Println("Number is odd")
    }
}
// Output: Number is odd
```

In this example, number := 7 is the initialization statement. The program then checks if number is even or odd, and prints the corresponding output.

3.1.3 Else-If Statement

The else-if statement in Go adds more flexibility to the basic if-else structure by allowing for multiple conditions to be checked in sequence. It is used when you have several different conditions and need to execute different code blocks depending on which condition is true. The else-if statement can be seen as a way to handle multiple decision-making processes within a single if-else construct.

Syntax

The syntax for using else-if in Go is as follows:

```
if condition1 {
    // code to execute if condition1 is true
} else if condition2 {
    // code to execute if condition2 is true
```

```
} else {
   // code to execute if none of the above conditions are true
}
```

Here, condition1 and condition2 are boolean expressions. The code in the else if block only executes if its condition is true and all previous conditions are false. If none of the conditions are true, the code in the else block is executed.

Features of the Else-If Statement in Go

- *Multiple Conditions:* Allows checking multiple conditions in a single if-else structure.
- *Sequential Checking:* Each condition is checked in order. Once a true condition is found, the corresponding code block is executed, and the rest of the else-if ladder is skipped.
- *Fallback with Else:* An else block can be used as a fallback, executing only if none of the else-if conditions are met.

Example

Let's look at an example to illustrate the use of else-if in Go:

```
package main
import "fmt"
func main() {
   score := 75
   if score >= 90 {
      fmt.Println("Grade: A")
   } else if score >= 80 {
```

```
        fmt.Println("Grade: B")
    } else if score >= 70 {
        fmt.Println("Grade: C")
    } else {
        fmt.Println("Grade: F")
    }
}
// Output: Grade: C
```

In this example, the program assigns a grade based on a score. The else-if ladder checks the score against various ranges and prints the corresponding grade. Here, since the score is 75, it falls into the else if score >= 70 condition, and "Grade: C" is printed.

3.2 Loops

Loops are fundamental constructs in programming that allow for the repeated execution of a block of code as long as a specified condition is met. They are essential for automating repetitive tasks, processing collections of data, and building complex programming structures. By using loops, programmers can significantly reduce the amount of code they write, enhance code readability, and manage complexity.

- *For Loop:* The most common loop, used for iterating over a range, collection, or series of elements. It repeats a block of code a certain number of times, based on a counter or an iterator.
- *While Loop:* This loop continues executing a block of code as long as its condition remains true. It's typically used when the number of iterations is not known beforehand.
- *Do-While Loop:* Similar to the while loop, but it guarantees that the code block is executed at least once, as the condition is checked after the execution of the block.

- *Infinite Loop:* A loop that runs indefinitely because the loop condition is always true. These are used intentionally in some scenarios but can also be the result of a programming error.
- *Nested Loops:* These involve placing one loop inside another loop. Nested loops are used for working with multi-dimensional data structures or performing more complex tasks that require a combination of loops.

3.2.1 For Loop

The for loop in Go is a versatile and powerful control structure that is used to repeat a block of code multiple times. It's the only looping construct in Go, but it's flexible enough to cover the functionality of traditional for, while, and do-while loops found in other languages.

Syntax

The basic syntax of a for loop in Go is as follows:

```
for initializer; condition; post {
   // code to be executed
}
```

initializer: This is executed before the first iteration, typically used to declare and initialize a loop counter.

condition: Before each iteration, this condition is evaluated. If true, the loop continues; if false, the loop terminates.

post: This is executed after each iteration, commonly used to update the loop counter.

Example

Here's an example of a traditional for loop in Go:

```
package main
import "fmt"
func main() {
   for i := 0; i < 5; i++ {
      fmt.Println("Loop iteration:", i)
   }
}
// Output:
// Loop iteration: 0
// Loop iteration: 1
// Loop iteration: 2
// Loop iteration: 3
// Loop iteration: 4
```

In this example, the loop starts with i := 0. The condition i < 5 is checked before each iteration, and the post statement i++ is executed after each iteration.

3.2.2 While-Style Loop

While Go does not have a specific while loop construct as found in many other programming languages, it achieves the same functionality through its versatile for loop. By omitting the initializer and post statement in a for loop, you effectively create a while-style loop in Go. This variant is used when you need to repeat a block of code as long as a condition remains true, but the number of iterations is not known beforehand.

Syntax

The syntax of a while-style loop in Go is a simplified version of the for loop:

```
for condition {
    // code to be executed as long as condition is true
}
```

Here, condition is a boolean expression. The loop continues executing the block of code as long as this condition evaluates to true.

Features of the While-Style Loop

- Simplicity: It offers a clean and clear way to execute a loop when only a condition is needed to control the loop.
- Flexibility: The condition can be any boolean expression, giving you flexibility in how you control the loop.

Example

Let's look at an example to illustrate the use of a while-style loop in Go:

```
package main
import (
    "fmt"
    "time"
)
func main() {
    timeout := time.After(5 * time.Second)
    tick := time.Tick(1 * time.Second)

    for {
        select {
```

```
        case <-timeout:
            fmt.Println("Timed out!")
            return
        case t := <-tick:
            fmt.Println("Tick at", t)
        }
    }
}
// Output:
// Tick at 2022-03-10 10:00:01 +0000 UTC
// Tick at 2022-03-10 10:00:02 +0000 UTC
// ...
// Timed out!
```

In this example, the program prints a message every second. After 5 seconds, it prints "Timed out!" and exits the loop. The loop keeps running until the timeout channel sends a signal.

3.2.3 Infinite Loop

In Go, an infinite loop is a loop that continues to execute indefinitely because the loop condition is always true or there is no condition specified. Infinite loops are a fundamental concept in programming, often used intentionally in scenarios where a program needs to run continuously until it's explicitly told to stop, such as in server processes or event listening.

Syntax

The syntax of an infinite loop in Go is a further simplification of the for loop, where you omit the loop condition:

```
for {
   // code to be executed continuously
}
```

In this construct, the loop lacks a terminating condition, which means the block of code inside the loop will execute repeatedly, without end.

Features of the Infinite Loop

No Termination Condition: The loop does not have a condition to check for termination; it will run forever unless externally interrupted.

Controlled Exit: Despite being infinite, you can still control the exit from the loop using break statements or returning from the loop's containing function.

Use in Event Loops: Commonly used in applications like web servers or GUIs where the program needs to wait for external events (like client requests or user actions).

Example

Here's an example of an infinite loop in Go:

```
package main
import (
   "fmt"
   "time"
)
func main() {
   for {
      fmt.Println("This loop will run forever...")
      time.Sleep(1 * time.Second)
```

```
    }
}
```

In this example, the program prints a message to the console every second indefinitely. The time.Sleep call is used to delay each iteration of the loop, making the output more readable and simulating a task that runs continuously.

Controlling Exit from an Infinite Loop

You can exit an infinite loop using a break statement. Here's an example:

```
package main
import (
    "fmt"
    "time"
)
func main() {
    start := time.Now()
    for {
        fmt.Println("Running...")
        time.Sleep(1 * time.Second)
        if time.Since(start) > 5*time.Second {
            fmt.Println("Exiting loop.")
            break
        }
    }
}
```

In this example, the loop runs for approximately 5 seconds. The if statement checks how much time has passed since the loop started and breaks out of the loop after 5 seconds.

Infinite loops in Go are powerful tools when used correctly. They're essential in situations where a program needs to run continuously, but they must be used with care to avoid unintentional infinite execution, which can lead to issues like unresponsiveness or excessive resource usage.

3.2.4 For-range Loop

The for-range loop in Go provides a convenient and powerful way to iterate over elements of various data structures such as slices, arrays, maps, strings, and channels. It allows you to access each element in these data structures without needing to manually manage an index or key, making your code cleaner and more readable.

Syntax

The basic syntax of a for-range loop in Go is as follows:

```
for index, value := range collection {
   // code to execute for each element in the collection
}
```

- index is the position of the current element in the collection. For maps, it's the key of the current element.
- value is the current element itself. This is a copy of the element for slices, arrays, and strings, and a key-value pair for maps.
- collection is the data structure you are iterating over.

If you don't need the index or key, you can ignore it using an underscore (_). Similarly, if you only need the index or key, you can omit the value.

Features of the For-Range Loop

- Automatic Iteration: Automatically handles the iteration over the collection without needing to specify the start and end conditions explicitly.
- Versatility: Works with several built-in data types, making it a versatile tool for data manipulation.
- Safety: Reduces the risk of off-by-one errors and other common mistakes associated with manual iteration.

Example

Iterating Over a Slice:

```
package main
import "fmt"
func main() {
    numbers := []int{1, 2, 3, 4, 5}
    for index, value := range numbers {
        fmt.Printf("Index: %d, Value: %d\n", index, value)
    }
}
```

This example iterates over a slice of integers, printing each index and its corresponding value.

Iterating Over a Map

```
package main
import "fmt"
func main() {
    fruits := map[string]int{"apple": 5, "banana": 3, "orange": 4}
```

```
    for key, value := range fruits {
        fmt.Printf("Key: %s, Value: %d\n", key, value)
    }
}
```

In this example, the for-range loop iterates over a map of fruits, where the key is the fruit's name and the value is the quantity. The loop prints each fruit and its quantity.

Iterating Over a String

```
package main
import "fmt"
func main() {
    for index, runeValue := range "hello" {
        fmt.Printf("Index: %d, Rune: %c\n", index,
runeValue)
    }
}
```

Here, the for-range loop iterates over a string, printing each rune's index and its value. Note that the index refers to the byte index, and the runeValue is the rune (character) at that index, which accounts for UTF-8 encoding.

The for-range loop in Go simplifies the process of iterating over collections and data structures, making it an essential feature for data manipulation and processing tasks in Go programming.

3.3 Switch Statements

Switch statements in Go provide a more streamlined and readable way to execute different blocks of code based on the value of a variable or an expression. They serve as an alternative to long chains of if-else-if statements, making the code easier to follow and manage. Switch statements evaluate an expression once and compare its value against multiple possible matches, executing the block of code associated with the first matching case. Following are various types of the switches in golang.

- **Expression Switches:** The most common type, where the switch statement evaluates an expression and compares its value to the cases specified in the switch block.
- **Type Switches:** Unique to Go, type switches compare the type of an interface value against a list of types. They are useful for type assertion in scenarios where the exact type of an interface value is unknown.
- **Tagless Switches:** Also known as "switch true" statements, these evaluate cases as boolean expressions. The first case to evaluate to true is executed. This form is particularly useful for simplifying complex conditionals.

Features of Switch Statements

- Simplifies Complex Conditionals: By replacing multiple if-else-if statements with a single switch, code becomes easier to read and understand.
- Implicit Breaks: In Go, each case in a switch statement ends with an implicit break, preventing fall-through by default. To explicitly continue to the next case, the fallthrough keyword must be used.
- Multiple Matches in a Single Case: Go allows multiple values to be specified in a single case, separated by commas, making it easy to group several values together.
- Flexibility: Go's switch statements can be used not only with constants but also with expressions and types, offering a high degree of flexibility in how conditions are evaluated.

3.3.1 Simple Switch

A simple switch statement:

```
switch variable {
case value1:
    // code to execute if variable == value1
case value2:
```

```
    // code to execute if variable == value2
default:
    // code to execute if no cases match
}
```

Example

```
switch day {
case "Monday":
    fmt.Println("Start of the week")
case "Friday":
    fmt.Println("Weekend is near")
default:
    fmt.Println("A regular day")
}
```

3.3.2 Switch with Initialization

You can also initialize a value in the switch statement itself.

```
switch initialization; condition {
case condition1:
    // code for condition1
case condition2:
    // code for condition2
default:
    // default code
}
```

Example

```
switch x := computeValue(); x {
case 1:
    fmt.Println("One")
default:
    fmt.Println("Other")
}
```

3.3.3 Tagless Switch

Go allows tagless switch statements, where cases contain boolean expressions.

```
switch {
case condition1:
    // code for condition1
case condition2:
    // code for condition2
default:
    // default code
}
```

Example

```
switch {
case x < 0:
    fmt.Println("Negative")
```

```
case x == 0:
    fmt.Println("Zero")
default:
    fmt.Println("Positive")
}
```

Bringing it All Together: The Recipe for Success

Let's revisit the cake-baking analogy we discussed at the beginning of this chapter. By now, you've learned about various control structures such as if-else statements, loops, and sequencing in Go. Here's how they correspond to the steps in baking a cake, showcasing their practical utility in programming:

- **Decision-making (if-else statements):** This is like deciding whether to buy missing ingredients or modify the recipe based on what's available in your pantry. In programming, if-else statements help you handle decisions within your code, allowing different actions based on specific conditions.
- **Repeating actions (loops):** Just as you might mix ingredients until they are just right or check the cake multiple times during baking, loops in programming allow you to repeat certain operations until a condition is met or no longer holds.
- **Sequencing:** Following the recipe steps in the right order ensures that your cake turns out as expected. Similarly, the correct sequence of code execution in programming ensures that your program functions correctly.

Just as you control the flow of the cake-baking process based on various conditions and outcomes, control structures in Go empower you to dictate how and when different pieces of code are executed. This ability makes your programs dynamic and responsive, capable of handling a variety of tasks and conditions with ease. Understanding and using these structures effectively is key to writing efficient and powerful code.

Chapter 4
Arrays and Slices in Go

In the world of Go programming, managing collections of data efficiently and effectively is crucial for building robust applications. Two fundamental data structures that serve this purpose are arrays and slices. Understanding these structures is key to mastering Go, as they form the backbone for handling sequences of data. Arrays and slices enable programmers to organize data in a structured way, making it easier to perform operations like sorting, searching, and manipulating lists of items.

Imagine you are organizing a bookshelf. Instead of placing books randomly, you might group them by genre, author, or publication year. In programming, arrays and slices help you manage similar kinds of data—such as user information, product details, or computational data—by storing them in an ordered manner. This organization allows you to access and modify your data efficiently.

This chapter will delve into arrays and slices, exploring their characteristics, differences, and how to use them in your Go programs. You'll learn how arrays provide a fixed-size container for elements, making them ideal for situations where the number of elements is known and unlikely to change. On the other hand, slices are dynamic and can grow or shrink, offering flexibility when dealing with data whose size might vary during runtime. By understanding how to leverage these structures, you'll be able to handle data more effectively, leading to more reliable and responsive applications.

Arrays

Arrays are one of the simplest data structures and a foundational element in Go programming. An array is a fixed-size sequence of elements of a single type. Once declared, the size of an array cannot be changed, making arrays a stable and efficient structure for managing a known quantity of elements. We'll cover how to declare arrays, initialize them, and access their elements. Additionally, we'll look at how arrays are passed to functions and how they can be used in multidimensional configurations for more complex data management tasks.

Slices

Slices, on the other hand, are more dynamic and flexible than arrays. They are built on top of arrays to provide a more powerful and convenient interface for working with sequences of data. A slice can grow and shrink, making it ideal for situations where the size of the collection is not known upfront or can change over time. We will explore slice creation, how to manipulate slices with built-in functions like append and copy, and understand the underlying mechanics that allow slices to be so versatile yet efficient. Slices are a critical part of Go's approach to data handling, providing a rich set of features that support a wide range of programming needs.

Comparing Arrays and Slices

While arrays and slices are closely related, understanding their differences is key to choosing the right one for your specific scenario. We'll compare and contrast these two data structures, highlighting when and why you might choose one over the other. Through examples and explanations, we'll demystify the most common operations and techniques for working with arrays and slices, ensuring you have a solid foundation for more advanced topics in Go programming.

This chapter aims to equip you with the knowledge and skills to effectively work with arrays and slices in Go. Whether you're managing fixed-size collections of elements or dealing with dynamic datasets that change over time, mastering arrays and slices will enhance your ability to write efficient, readable, and powerful Go code.

4.1 Arrays

Arrays are a fundamental data structure in Go that allow you to store fixed-size collections of elements of the same type. They play a crucial role in organizing data efficiently, enabling quick access and manipulation of stored elements. Understanding how to create and use arrays is essential for any Go programmer.

4.1.1 Importance of Arrays

Arrays are important because they provide a way to group related data items together under a common name. This organization facilitates data management by allowing you to iterate over elements for processing and perform operations on each element using a single identifier. Arrays also have a fixed size, which means the amount of memory they occupy is known at compile time, leading to efficient memory usage.

4.1.2 Usage

Arrays are used in scenarios where the number of elements is known beforehand, and the data collection size is not expected to change. They are particularly useful for applications that require fast access to elements, as arrays provide constant-time access to any element via its index. Common use cases include storing fixed collections of items, such as configuration settings, lookup tables, or temporary data for processing in loops. Arrays form the basis for more complex data structures like slices, which offer more flexibility for dynamic data collections. Understanding arrays and their characteristics is a stepping stone to mastering slices and effectively managing collections of data in Go.

Syntax

```
var arrayName [size]ElementType
```

Here,

- arrayName is the identifier for the array.
- size is a non-negative integer that defines the number of elements the array can hold.
- ElementType is the type of elements stored in the array.

Example

Here's an example of how to create and initialize an array in Go:

```
package main
import "fmt"
func main() {
   // Declare an array of 5 integers
   var numbers [5]int
   // Initialize the array elements
   numbers[0] = 2
   numbers[1] = 4
   numbers[2] = 6
   numbers[3] = 8
   numbers[4] = 10
   // Print the array elements
   fmt.Println(numbers) // Output: [2 4 6 8 10]
}
```

In this example, an array named numbers is created to hold 5 integers. Each element in the array is then assigned a value, and finally, the contents of the array are printed.

4.1.3 Manipulating Arrays

Manipulating arrays in Go involves changing, adding, or removing elements. However, due to their fixed size, arrays in Go have limited manipulation capabilities compared to slices. The primary way to manipulate an array is by changing the value of its elements.

Syntax

```
arrayName[index] = newValue
```

Here,

- arrayName is the identifier of the array.
- index is the position of the element you want to change, starting from 0.
- newValue is the new value you want to assign to the element.

Example

```
package main
import "fmt"
func main() {
    var numbers [5]int = [5]int{1, 2, 3, 4, 5}
    // Changing the value of the third element
    numbers[2] = 10
    fmt.Println(numbers) // Output: [1 2 10 4 5]
}
```

In this example, we change the third element of the numbers array from 3 to 10.

Concerns with Manipulating Arrays

- Fixed Size: Arrays cannot grow or shrink, limiting the type of manipulations you can perform.
- Value Semantics: When you assign an array to another variable or pass it to a function, Go copies the entire array. This behavior can lead to performance issues with large arrays.

4.2 Slices

Slices in Go are a dynamic, flexible alternative to arrays, providing powerful capabilities for working with sequences of data. Unlike arrays, slices do not have a fixed size, making them ideal for situations where the collection of data is either not known at compile time or needs to change size during execution.

4.2.1 Importance of Slices

Slices are critically important in Go due to their dynamic nature. They use an underlying array to manage data but expose a more flexible interface to the developer. This flexibility allows for more efficient and convenient data manipulation, such as appending elements, slicing subsets of data, and dynamically resizing the collection. Slices are a key component of Go's approach to data handling, especially when dealing with collections of data that must be modified or extended over time.

4.2.2 Usage

Slices are extensively used throughout Go for handling sequences of data. They are particularly useful in scenarios where the size of the data collection may change, such as:

- Reading lines from a file when the number of lines is unknown.
- Accumulating elements from user input or other dynamic sources.
- Manipulating subsets of data for processing or analysis.

Understanding slices and their operations is essential for effective Go programming, given their central role in data manipulation and Go's built-in support for slice operations. Slices provide a powerful, flexible way to work with collections of data, making them one of the most widely used features in Go.

Syntax

Slices can be created in several ways in Go:

- Using the make function: This is the most common way to create a slice with a predefined size.

```
sliceName := make([]ElementType, length, capacity)
```

Here,

- sliceName is the identifier for the slice.
- ElementType is the type of elements stored in the slice.
- length is the initial length of the slice (the number of elements it contains).
- capacity is optional and specifies the initial capacity of the underlying array.
- Slice literals: This method initializes the slice with specific

```
sliceName := []ElementType{value1, value2, value3}
```

Example

Here's an example of creating and using a slice in Go:

```
package main
import "fmt"
func main() {
    // Creating a slice using a slice literal
    fruits := []string{"apple", "banana", "cherry"}
    // Appending an element to the slice
    fruits = append(fruits, "date")
    // Slicing a subset
    subFruits := fruits[1:3]
    // Printing slices
    fmt.Println(fruits)    // Output: [apple banana cherry date]
    fmt.Println(subFruits) // Output: [banana cherry]
```

```
}
```

4.2.3 Manipulating Slices

Slices are much more flexible than arrays, providing a variety of ways to manipulate the collection of data they represent. Operations include appending elements, slicing to create sub-slices, and copying elements between slices.

Appending to a Slice

The append function is used to add new elements to the end of a slice:

```
sliceName = append(sliceName, newElement)
```

You can also append multiple elements or another slice:

```
sliceName = append(sliceName, elem1, elem2)
sliceName = append(sliceName, anotherSlice...)
```

Example

```
package main
import "fmt"
func main() {
    fruits := []string{"apple", "banana"}
    fruits = append(fruits, "cherry", "date")
    fmt.Println(fruits) // Output: [apple banana cherry date]
}
```

In this example, two elements are appended to the fruits slice.

Slicing to Create Sub-slices

You can create a sub-slice from a slice by specifying a low and high bound:

Syntax

```
subSlice := sliceName[low:high]
```

Example

```
package main
import "fmt"
func main() {
   numbers := []int{1, 2, 3, 4, 5}
   subNumbers := numbers[1:4] // This will include elements at index 1, 2, and
3
   fmt.Println(subNumbers)
}
Output:
[2 3 4]
```

Copying Elements Between Slices

The copy function copies elements from one slice to another:

Syntax

```
copy(destSlice, srcSlice)
```

Example

In this example, elements from src are copied to dest. Since dest can only hold 2 elements, only the first two elements of src are copied.

```
package main
import "fmt"
func main() {
   src := []int{1, 2, 3}
   dest := make([]int, 2)
   copy(dest, src)
   fmt.Println(dest)
}
Output:
[1 2]
```

Concerns with Manipulating Slices

- Capacity and Performance: While appending, if the underlying array of a slice does not have enough capacity to accommodate new elements, a new array is allocated, and elements are copied over. This can impact performance.
- Aliasing: When you create a sub-slice, both the original slice and the sub-slice refer to the same array. Modifying elements via one slice affects the other.

4.3 Multi Dimensional Arrays

Multi-dimensional arrays in Go are arrays that contain other arrays as their elements. These structures are used to organize and manipulate data in a tabular form or any scenario requiring a grid-like layout, such as matrices. Understanding how to work with multi-dimensional arrays is crucial for solving complex problems that involve nested data structures.

A multi-dimensional array can be thought of as an "array of arrays". The most common type is the two-dimensional array, but Go supports arrays with more dimensions as needed. Each level of a multi-dimensional array adds another layer of indexing, allowing for more complex data relationships and structures.

Use Cases

Multi-dimensional arrays are particularly useful in scenarios where you need to store and manipulate data in a structured format. Common use cases include:

- Matrices and Linear Algebra: Operations on matrices, such as addition, multiplication, or transposition, are fundamental in many areas of computing, including graphics, physics simulations, and machine learning.
- Game Development: Representing game boards (like chess or tic-tac-toe) or mapping environments where elements are positioned in a grid.
- Data Analysis: Organizing and analyzing data in tabular formats, such as spreadsheets or relational database tables.

Real-world Examples

Chess Game Board: In a chess game, the board can be represented as a two-dimensional array where each element corresponds to a square that may contain a piece. The rows and columns represent the ranks and files of the chessboard, respectively.

Pixel-based Images: Image data, especially in black and white or grayscale, can be represented using a two-dimensional array. Each element of the array would correspond to a pixel's intensity value, with rows and columns representing the image's dimensions.

Syntax

The syntax for declaring a multi-dimensional array in Go is as follows:

```
var arrayName [size1][size2]ElementType
```

Here,

- size1 and size2 represent the dimensions of the array.
- ElementType is the type of elements stored in the array.

Example

In this example, a two-dimensional array matrix is defined to represent a simple 2x3 matrix. Each element is then initialized, and the matrix is printed.

Multi-dimensional arrays are a powerful tool in Go for organizing complex, structured data. While they are a bit more complicated to manage than single-dimensional arrays, their ability to represent data in multiple dimensions makes them indispensable for certain types of problems and applications.

```
package main
import "fmt"
func main() {
    // Declare a 2x3 two-dimensional integer array
    var matrix [2][3]int
    // Initialize array elements
    matrix[0][0] = 1
    matrix[0][1] = 2
    matrix[0][2] = 3
    matrix[1][0] = 4
    matrix[1][1] = 5
    matrix[1][2] = 6
    // Print the two-dimensional array
    fmt.Println("Matrix:", matrix)
    // Output: Matrix: [[1 2 3] [4 5 6]]
}
```

Chapter 5
Pointers and Memory Management

This chapter delves into the foundational concepts of pointers and memory addresses, critical components in Go that allow direct interaction with memory locations. Understanding pointers is essential for optimizing your programs' performance and resource management. As Go handles memory allocation and deallocation through an efficient garbage collection system, mastering these concepts enables you to write cleaner, more efficient code.

Imagine being able to control exactly where your program stores its data and how it accesses this data. Pointers give you this control, allowing precise manipulation of memory, which is especially useful in systems programming, data structures, and when interfacing with hardware. Memory management is a pivotal aspect of programming in Go, as it directly impacts the performance and scalability of applications.

In this chapter, we will explore how Go manages memory through pointers and addresses. You'll learn about the role of garbage collection in automating memory management, which helps prevent memory leaks and other common errors associated with manual memory handling. By the end of this chapter, you will have a thorough understanding of how Go's memory management works, enabling you to develop applications that make better use of system resources and deliver superior performance.

5.1 Understanding Pointers and Addresses

Every variable resides at a specific location in memory. This location is uniquely identified by a memory address, which can be thought of as the variable's "home address" within the vast landscape of your computer's memory.

A memory address serves as a direct reference to the physical location where data is stored in RAM (Random Access Memory). In Go, when you create a variable, the Go runtime allocates a space in memory for that variable and assigns it a memory address. This address is used internally to access the variable's value efficiently.

Syntax

The syntax for declaring a pointer in Go is to prefix the type with an asterisk *. For example, a pointer to an int is declared as *int.

Creating Pointers: You create a pointer by using the & operator to find the address of a variable:

```
var x int = 10

var p *int = &x
```

Dereferencing Pointers: Dereferencing a pointer is done using the * operator, allowing access to the value at the pointer's address:

```
var value int = *p // value now holds the value of x, which is 10
```

5.2 Accessing Memory Addresses

To access the memory address of a variable in Go, you use the **&** operator placed before the variable's name. This operator fetches the memory address where the variable is stored. The resulting value is known as a "pointer," because it "points" to a location in memory.

Syntax

```
address := &variableName
```

Here,

- variableName is the identifier of the variable whose address you want to access.
- address will now hold the memory address of variableName.

Example

Consider a simple integer variable:

```
package main
import "fmt"
func main() {
    myNumber := 42
    fmt.Println("Value of myNumber:", myNumber)
    fmt.Println("Address of myNumber:", &myNumber)
}
Output:
Value of myNumber: 42
Address of myNumber: 0xc0000120b8 (The actual address may vary)
```

In this example, myNumber is an integer variable with a value of 42. Using &myNumber, we access and print the memory address where myNumber is stored. The printed address (e.g., 0xc0000120b8) represents the specific location in memory allocated for myNumber.

5.3 Pointers and Functions

In Go, passing variables or structures to functions by their pointer references is a powerful technique that allows functions to modify the original data rather than working on a copy. This method of passing arguments can lead to more efficient memory usage, especially for large structures or when aiming for in-place modifications without returning a value.

5.3.1 Passing Pointer References for Variables

When you pass a pointer to a variable to a function, the function receives the memory address of the variable. Any modifications made to the variable through the pointer within the function will reflect on the original variable because both the function parameter and the original variable point to the same memory location.

Syntax

```
func myFunction(param *MyType) {
   // Function implementation
}
```

Here,

- param is a pointer to a variable of type MyType.

Example

```
package main
import "fmt"
func increment(number *int) {
   *number += 1
}
func main() {
   myNumber := 10
   increment(&myNumber)
   fmt.Println(myNumber)
}
Output:
11
```

In this example, increment takes a pointer to an int as its parameter. By dereferencing the pointer (*number) and modifying its value, increment affects the original myNumber variable outside the function.

5.3.2 Passing Pointer References for Structures

Passing pointers to structures (structs) is particularly useful when you want to modify the struct or avoid copying it entirely during the function call, which can be costly for large structs.

Syntax

```
func myFunction(param *MyStruct) {
   // Function implementation
}
```

Here,

- param is a pointer to an instance of MyStruct.

Example

```
package main
import "fmt"
type Person struct {
   Name string
   Age  int
}
func birthday(p *Person) {
   p.Age += 1
}
func main() {
```

```
    john := Person{Name: "John", Age: 30}
    birthday(&john)
    fmt.Println(john) // Output: {John 31}
}
```

In this example, birthday takes a pointer to a Person struct. It increments the Age field of the struct it points to, effectively modifying the original john instance passed to it.

5.4 Memory Management

Go manages memory allocation and deallocation automatically. Memory for new variables is allocated with the new function or composite literals, and unused memory is reclaimed by the garbage collector.

5.4.1 Garbage Collection

One of Go's most significant features is its garbage collector (GC), which automatically reclaims memory occupied by objects that are no longer accessible by the program. This process runs concurrently with the application, working to free memory in the background without requiring explicit deallocation from the programmer.

How Garbage Collection Works

Go's garbage collector implements a form of "tricolor marking" algorithm, which conceptually divides objects into three categories:

- White: Objects that are potentially unreachable.
- Grey: Objects that are reachable but whose references haven't been scanned yet.
- Black: Objects that are reachable, and all references have been scanned.

The GC starts by considering all objects white. It then scans from root objects (globals, stack variables) and colors reachable objects grey, turning them black as it scans their references. Finally, any objects that remain white are considered unreachable and are collected.

Advantages of Go's Garbage Collection

- Safety: Automatic garbage collection helps prevent memory leaks and dangling pointer errors, common issues in manual memory management.
- Concurrency: The GC runs concurrently with the application, aiming to minimize pause times and reduce the impact on program performance.
- Simplicity: By abstracting memory deallocation, Go allows developers to focus on application logic rather than memory lifecycle management.

Chapter 6
Structs and Interfaces

Structs provide a way to group related data together into cohesive units, akin to custom types. This allows developers to model real-world entities and relationships in a clear and maintainable manner, much like creating a blueprint for an object in the real world with attributes such as size, shape, and functionalities. Structs serve as the cornerstone for building complex data structures, encapsulating the attributes of entities in your program with ease and precision.

On the other hand, **Interfaces** in Go elevate the concept of abstraction, defining sets of behaviors that structs or any other types can implement. This level of abstraction allows for more flexible and decoupled design patterns, enabling developers to write functions and methods that can operate on any type that satisfies the interface. The beauty of interfaces lies in their ability to promote a design that focuses on what actions an object can perform rather than strictly what it is. This approach leads to code that is both extensible and resilient to changes, facilitating the development of large-scale, modular applications.

Imagine you are constructing a digital city. Structs would be like the different buildings, each designed for specific purposes—residential buildings, offices, schools, etc., each with its own design and structure. Interfaces would be like the regulations that all buildings must follow, such as safety codes and functionality requirements, allowing city planners (developers) to ensure that regardless of the building type, certain standards are met.

By understanding and utilizing structs and interfaces effectively, you can architect your Go applications to be robust, scalable, and adaptable. This chapter will guide you through the nuances of these powerful features, illustrating how they can be used to build well-organized and efficient software solutions.

6.1 Structs

Structs in Go are composite data types that group together variables under a single name. These variables, known as fields, can be of different types, making structs incredibly versatile for data organization. Structs are particularly useful for modeling complex data, representing objects with properties, or grouping data together for passing around in your programs.

Syntax

The basic syntax for declaring a struct in Go involves the type keyword, followed by the name of the struct and the struct keyword. Inside the curly braces, you list the fields with their names and types.

```
type StructName struct {
    Field1 Type1
    Field2 Type2
    // add more fields as needed
}
```

Here,

- StructName is the name of the struct.
- Field1, Field2, etc., are the names of the fields within the struct.
- Type1, Type2, etc., indicate the data types of each field.

6.1.1 Instantiating Structs

Creating instances of structs is a fundamental operation in Go programming, enabling you to work with concrete data that adheres to the templates defined by your struct types. Go offers several methods for instantiating structs, each with its own use cases and implications for memory management and access patterns. Understanding these methods is crucial for effectively utilizing structs in your Go applications.

Using the new Keyword

The new function in Go allocates memory for a struct, initializing each field to its zero value, and returns a pointer to the newly allocated struct. This method is useful when you want to work with pointers to structs right from the start.

Syntax

```
instancePtr := new(MyStruct)
```

Here,

- MyStruct is the struct type.
- instancePtr is a pointer to an instance of MyStruct, with each field initialized to its zero value.

Example

```
package main
import "fmt"
type Person struct {
   Name string
   Age  int
}
```

```
func main() {
    personPtr := new(Person)
    personPtr.Name = "Alice"
    personPtr.Age = 30
    fmt.Println(*personPtr) // Output: {Alice 30}
}
```

In this example, new(Person) creates a new instance of the Person struct, with Name and Age initialized to their zero values ("" and 0, respectively). We then set the fields and print the dereferenced struct.

Using Composite Literals

Composite literals provide a concise syntax for creating struct instances, allowing you to specify initial values for fields at the time of creation. This method can be used to create both pointer and value instances.

Syntax

- Value instance: instance := MyStruct{Field1: value1, Field2: value2}
- Pointer instance: instancePtr := &MyStruct{Field1: value1, Field2: value2}

Example

```
package main
import "fmt"
type Person struct {
    Name string
    Age  int
}
```

```
func main() {
  // Creating a value instance with composite literals
  alice := Person{Name: "Alice", Age: 30}
  // Creating a pointer instance with composite literals
  bobPtr := &Person{Name: "Bob", Age: 25}
  fmt.Println(alice)  // Output: {Alice 30}
  fmt.Println(*bobPtr) // Output: {Bob 25}
}
```

Differences Between Pointer and Value Instances

- Memory Semantics: Pointer instances allow for shared access to a single struct in memory. Modifications through one pointer are visible to all references of that struct. Value instances, on the other hand, create copies of the struct; modifications to a copy do not affect the original struct or other copies.
- Zero Value Initialization: Using new or taking the address of a composite literal (&MyStruct{...}) initializes fields to zero values if not explicitly specified. Composite literals without & provide direct initialization of fields and do not necessarily require all fields to be mentioned.

6.1.2 Initializing Structs

Structs can be initialized with zero values, with specific field values directly at creation, or through the use of struct tags which can significantly enhance functionality, particularly in serialization and deserialization processes.

Initializing Structs with Zero Values

By default, when a struct is instantiated without explicitly providing initial values, each field is initialized to its zero value (e.g., 0 for int, "" for string, false for bool, nil for pointers, slices, maps, channels, and functions).

Syntax

```
var myStruct MyStruct
```

Example

```
package main
import "fmt"
type Person struct {
   Name string
   Age  int
}
func main() {
   var person Person
   fmt.Println(person) // Output: { 0}
}
```

Initializing Structs with Specific Field Values

Go allows you to initialize structs with specific values using composite literals. This is a powerful feature that lets you set initial values for some or all of the struct's fields.

Syntax

```
myStruct := MyStruct{Field1: value1, Field2: value2}
```

This syntax initializes myStruct with value1 and value2 for Field1 and Field2, respectively. It's also possible to create a pointer to a struct instance directly:

Example

```
package main
import "fmt"
type Person struct {
    Name string
    Age  int
}
func main() {
    alice := Person{Name: "Alice", Age: 30}
    fmt.Println(alice) // Output: {Alice 30}
}
```

Here, alice is explicitly initialized with Name as "Alice" and Age as 30, demonstrating how to set specific field values at the time of struct creation.

The Role of Struct Tags

Struct tags in Go provide a way to attach metadata to the fields of a struct, commonly used with packages like encoding/json for controlling serialization and deserialization behavior.

Syntax

```
type MyStruct struct {
    Field1 Type1 `tag_name:"tag_value"`
    Field2 Type2 `tag_name:"tag_value"`
}
```

Struct tags are defined within backticks following the type declaration of a field.

Example

```
package main
import (
    "encoding/json"
    "fmt"
)
type Person struct {
    Name string `json:"name"`
    Age  int    `json:"age,omitempty"`
}
func main() {
    // Deserializing JSON into a Person struct
    jsonStr := `{"name": "Bob", "age": 25}`
    var person Person
    json.Unmarshal([]byte(jsonStr), &person)
    fmt.Println(person) // Output: {Bob 25}
    // Serializing a Person struct to JSON, omitting zero value of age
    youngPerson := Person{Name: "Eve"}
    youngPersonJSON, _ := json.Marshal(youngPerson)
    fmt.Println(string(youngPersonJSON)) // Output: {"name":"Eve"}
}
```

In this example, struct tags are used to map struct fields to JSON keys. The json:"name" tag maps the Name field to the "name" JSON key, and json:"age,omitempty" tells the json package to omit the Age field if it has a zero value during serialization.

6.1.3 Accessing and Modifying Struct Fields

Structs provide a way to group related data together, and being able to interact with this data is crucial for building functional and efficient applications. Additionally, understanding how to work with struct pointers is essential for effective memory management and performance optimization.

Accessing Struct Fields

To access a field of a struct, you use the dot (.) operator followed by the name of the field. This syntax is straightforward when you have a struct value.

Syntax

```
structVariable.FieldName
```

Here,

- structVariable is an instance of your struct.
- FieldName is the name of the field you wish to access.

Example

```
package main
import "fmt"
type Person struct {
   Name string
   Age  int
}
func main() {
```

```
    person := Person{Name: "Alice", Age: 30}
    fmt.Println(person.Name) // Output: Alice
    fmt.Println(person.Age)  // Output: 30
}
```

Modifying Struct Fields

Modifying the fields of a struct is also done using the dot operator. You simply assign a new value to the field you wish to change.

Syntax

```
structVariable.FieldName = newValue
```

Example

```
package main
import "fmt"
type Person struct {
    Name string
    Age  int
}
func main() {
    person := Person{Name: "Alice", Age: 30}
    person.Name = "Bob" // Modifying the Name field
    fmt.Println(person.Name) // Output: Bob
}
```

Here, we modify the Name field of the person struct to "Bob".

Accessing and Modifying Fields Through a Struct Pointer

When you have a pointer to a struct, you can still access and modify its fields without explicitly dereferencing the pointer. Go allows you to use the dot operator directly on a struct pointer, and it automatically dereferences the pointer for you.

Syntax

```
structPointer.FieldName
```

Here,

- structPointer is a pointer to your struct instance.
- Even though structPointer is a pointer, you use the same dot operator to access or modify its fields.

Example

```
package main
import "fmt"
type Person struct {
    Name string
    Age  int
}
func main() {
    personPtr := &Person{Name: "Alice", Age: 30}
    personPtr.Name = "Bob" // Automatically dereferenced
    fmt.Println(personPtr.Name) // Output: Bob
}
```

6.1.4 Struct Embedding and Composition

Struct embedding is achieved by declaring a field in a struct with the type of another struct but without giving it a name. The fields of the embedded struct become accessible through the outer struct, effectively allowing the outer struct to "inherit" the fields and methods of the inner struct. This mechanism promotes code reuse and simplifies the aggregation of behaviors. This feature allows one struct to include another struct as if its fields were part of the outer struct. This approach fosters composition over inheritance, emphasizing the construction of complex types by combining simpler, reusable ones.

Syntax

To embed one struct within another, simply declare the embedded struct type within the enclosing struct without providing a field name:

```
type EmbeddedStruct struct {
    // fields and methods
}
type OuterStruct struct {
    EmbeddedStruct
    // other fields and methods
}
```

Example

```
package main
import "fmt"
// Address struct represents a generic address
type Address struct {
    City, State string
```

```
}
// Person struct embeds Address, inheriting its fields
type Person struct {
    Name string
    Address
}
func main() {
    person := Person{
        Name: "Alice",
        Address: Address{
            City:  "Seattle",
            State: "WA",
        },
    }
    // Accessing embedded fields directly through the outer struct
    fmt.Println(person.Name)    // Output: Alice
    fmt.Println(person.City)   // Output: Seattle
    fmt.Println(person.State)  // Output: WA
}
```

In this example, Person embeds Address, allowing Person instances to directly access the fields of Address as if they were their own. This demonstrates how embedding facilitates a form of inheritance, enabling Person to compose its behavior and data from Address.

Accessing Embedded Struct's Methods

Embedded structs not only share their fields with the embedding struct but also their methods. This means that methods defined on the Address struct, for example, can be called on a Person instance directly, as long as Person embeds Address.

```
package main
import "fmt"
type Address struct {
   City, State string
}
// A method on Address
func (a Address) FullAddress() string {
   return fmt.Sprintf("%s, %s", a.City, a.State)
}
type Person struct {
   Name   string
   Address
}
func main() {
   person := Person{
      Name: "Alice",
      Address: Address{
         City:  "Seattle",
         State: "WA",
      },
```

```
    }
    // Calling a method of the embedded Address struct directly on Person
    fmt.Println(person.FullAddress()) // Output: Seattle, WA
}
```

6.1.5 Methods on Structs

In Go, methods are functions that are associated with a particular type. Defining methods on structs is a way to give behavior to the structs, allowing them to act on the data they encapsulate. Methods can be defined with either value receivers or pointer receivers, affecting how modifications to the struct fields within the methods are handled and applied.

Defining Methods on Structs

To define a method on a struct, you specify the receiver type between the func keyword and the method name. The receiver can be either a value receiver or a pointer receiver, depending on whether you need to modify the struct's fields within the method.

Syntax

```
func (receiver ReceiverType) MethodName(params) returnType {
    // Method body
}
```

Here,

- receiver is the instance of the struct the method is called on.
- ReceiverType is the type for which the method is defined. It can be a struct type or a pointer to a struct type.
- MethodName is the name of the method.

- params and returnType are the method's parameters and return type, respectively.

Value Receiver vs. Pointer Receiver Methods

- Value Receiver: Creates a copy of the struct for use within the method. Use a value receiver when the method does not need to modify the struct's fields or when you want to ensure the method operates on a copy of the struct, not the struct itself.

Example

```
package main
import "fmt"
type Circle struct {
   Radius float64
}
// Area method has a value receiver of type Circle
func (c Circle) Area() float64 {
   return 3.14 * c.Radius * c.Radius
}
func main() {
   circle := Circle{Radius: 5}
   fmt.Println(circle.Area()) // Output: 78.5
}
```

In this example, the Area method has a value receiver, meaning it receives a copy of the Circle instance. Modifications to c within Area would not affect the original circle instance.

- Pointer Receiver: Passes a pointer to the struct, allowing the method to modify the original struct's fields. Use a pointer receiver when the method needs to change the struct's state or when passing the struct by value would be inefficient (e.g., for large structs).

Example

```
package main
import "fmt"
type Circle struct {
   Radius float64
}
// Scale method has a pointer receiver of type *Circle
func (c *Circle) Scale(factor float64) {
   c.Radius = c.Radius * factor
}
func main() {
   circle := Circle{Radius: 5}
   circle.Scale(2)
   fmt.Println(circle.Radius) // Output: 10
}
```

In this case, the Scale method uses a pointer receiver, allowing it to modify the Radius field of the Circle instance. The change is reflected in the original circle instance because the method operates on a pointer to a circle, not a copy.

6.1.6 Exported vs. Unexported Struct Fields

Go's approach to encapsulation and visibility of struct fields hinges on the case of the first letter of the field names. This simple yet effective rule governs how struct fields can be accessed from outside the package in which they are defined. Understanding the distinction between exported and unexported struct fields is crucial for designing Go packages that are secure, maintainable, and properly encapsulated.

Visibility Rules

- Exported Fields: A struct field is exported (visible outside its package) if its name begins with an uppercase letter. Exported fields can be accessed from any package that imports the package where the struct is defined.
- Unexported Fields: A struct field is unexported (not visible outside its package) if its name begins with a lowercase letter. Unexported fields are intended for internal use within the package where the struct is defined and cannot be directly accessed from outside the package.

Syntax

```
package mypackage
// MyStruct is an example struct with both exported and unexported fields
type MyStruct struct {
    ExportedField   string // Exported: accessible outside the package
    unexportedField int    // Unexported: not accessible outside the package
}
```

In this example, ExportedField is accessible from any other package that imports mypackage,

6.2 Interfaces

Interfaces in Go are a powerful feature that define a contract for behavior. An interface specifies a set of method signatures (behavior) but does not implement them. Instead, types (usually structs) implement these methods, thereby fulfilling the interface. Interfaces enable polymorphism in Go, allowing you to write flexible, modular code that can work with any type that implements the interface's methods.

Encouraging Decoupling

One of the primary benefits of using interfaces is the decoupling they provide. By defining a contract that various types can satisfy, interfaces allow different parts of a program to communicate without being tightly bound to specific implementations. This decoupling means that changes in one part of the system (such as the implementation of a service) are less likely to require changes in the parts of the system that use that service, as long as the interface contract remains satisfied. Consequently, interfaces make the codebase more resilient to changes and easier to maintain.

Facilitating Code Reusability

Interfaces abstract away the specifics of implementation, enabling the same piece of code to work with various data types and structures. This polymorphism ensures that functions or methods that accept interfaces as parameters can operate on any value that implements the interface. As a result, developers can write more generic, reusable code libraries that can serve a broader set of applications, enhancing the code's reusability across different parts of a project or even across different projects.

Simplifying Testing and Mocking

In testing, particularly unit testing, interfaces play a crucial role by allowing for easy mocking of dependencies. Since interfaces define behavior without specifying the underlying implementation, testers can create mock objects that implement the same interfaces as the application's real objects. These mock objects can simulate various scenarios, making it easier to test components in isolation and ensuring that tests are not brittle or overly dependent on the concrete implementations of their dependencies.

Supporting Clean Architecture

Interfaces are instrumental in implementing clean architecture principles, where the emphasis is on separating concerns and making distinctions between the core business logic and the application's more fluid parts, such as UI, databases, and external services. By defining interfaces at the boundaries between these layers, developers can build systems that are more adaptable to change, whether due to new business requirements or the integration of new technologies.

Enhancing Code Readability and Maintenance

Code that leverages interfaces tends to be more abstract and focused on what actions are being performed rather than how they are implemented. This abstraction can lead to cleaner, more readable code that emphasizes the system's high-level behavior. Moreover, when interfaces are used consistently, they provide a clear and consistent pattern that developers can quickly understand and follow, reducing the cognitive load required to work with the codebase and facilitating easier maintenance and extension of the system over time.

6.2.1 Defining Interfaces

The syntax to declare an interface in Go is straightforward. It uses the type keyword followed by the interface name and the interface keyword. Inside the curly braces {}, you list the method signatures that any implementing type must have. These methods are defined without their bodies, as interfaces only describe behavior rather than implement it.

Syntax

```
type MyInterface interface {
    Method1(paramType1, paramType2) returnType1
    Method2(paramType) returnType2
    // Additional methods can be added here
}
```

Here,

- MyInterface is the name of the interface.
- Method1, Method2, etc., are the methods that the interface requires.

Example

```
package main
import "fmt"
// Greeter interface with a single method
type Greeter interface {
   Greet() string
}
// Person struct that will implement Greeter
type Person struct {
   Name string
}
// Greet method implemented by Person
// This method makes Person an implementation of the Greeter interface
func (p Person) Greet() string {
   return "Hello, my name is " + p.Name
}
func main() {
   var g Greeter = Person{Name: "Alice"}
   fmt.Println(g.Greet()) // Output: Hello, my name is Alice
}
```

In this example, the Greeter interface requires a Greet method that returns a string. The Person struct implements this method, thus implicitly implementing the Greeter interface. By defining the Greet method on Person, we fulfill the contract established by the Greeter interface.

6.2.2 Interfaces in functions

One of the most powerful features of interfaces in Go is their ability to allow functions to accept parameters of any type that implements the interface. This capability provides immense flexibility in function design and implementation, enabling the creation of generic, reusable code modules that can operate on a wide variety of types. By leveraging interfaces in function signatures, developers can write functions that are decoupled from the specifics of the types they operate on, focusing instead on the behavior those types exhibit.

Syntax

```
func FunctionName(param InterfaceName) ReturnType {
    // Function implementation
}
```

Here,

- FunctionName is the name of the function.
- param is the parameter that the function accepts, specified as an interface type.
- InterfaceName is the name of the interface that the parameter must implement.
- ReturnType is the type of value the function returns.

Example

```
package main
import "fmt"
// Speaker interface defines a method Speak
type Speaker interface {
```

```
    Speak() string
}
// Dog struct represents a type that can implement Speaker
type Dog struct {
    Name string
}
// Speak method implementation for Dog
func (d Dog) Speak() string {
    return fmt.Sprintf("Woof! I'm %s.", d.Name)
}
// Robot struct represents another type that can implement Speaker
type Robot struct {
    Model string
}
// Speak method implementation for Robot
func (r Robot) Speak() string {
    return fmt.Sprintf("Beep boop. I am model %s.", r.Model)
}
// Introduce takes a Speaker and calls its Speak method
func Introduce(s Speaker) {
    fmt.Println(s.Speak())
}
func main() {
    dog := Dog{Name: "Buddy"}
```

```
    robot := Robot{Model: "XJ-9"}
    Introduce(dog)   // Output: Woof! I'm Buddy.
    Introduce(robot) // Output: Beep boop. I am model XJ-9.
}
```

In this example, both Dog and Robot types implement the Speaker interface by providing their versions of the Speak method. The Introduce function, which takes a Speaker interface as its parameter, can accept any type that satisfies the Speaker interface. This demonstrates how interfaces allow functions to be written in a type-agnostic way, focusing on the behavior that types exhibit rather than the types themselves.

6.2.3 Empty Interface

In Go, the empty interface (interface{}) holds a special place. It is defined as an interface that specifies zero methods. Due to its definition, it can represent any type, making it an incredibly flexible tool for handling values of uncertain type. This characteristic of the empty interface opens up various possibilities for generic programming patterns and dynamic type handling in Go.

Syntax

The empty interface is specified as interface{}. Its primary feature is its ability to hold values of any type because every type, at the very least, implements zero methods. Here's the basic syntax:

```
var anyType interface{}
```

This declaration states that anyType is a variable that can hold a value of any type, thanks to the empty interface.

Example

```
package main
import "fmt"
func PrintDetails(v interface{}) {
   fmt.Printf("Value: %v, Type: %T\n", v, v)
}
func main() {
   PrintDetails("Hello, Go!")
   PrintDetails(42)
   PrintDetails(3.14)
}
```

In this example, PrintDetails accepts a parameter v of type interface{}, allowing it to handle values of any type. The function prints both the value and its type, showcasing the empty interface's ability to work with any type.

Implications of Using the Empty Interface

While the empty interface provides significant flexibility, its use comes with certain implications and considerations:

- Type Safety: Overuse of the empty interface can lead to code that loses the benefits of Go's strong typing system. Without explicit types, you risk runtime errors if you make incorrect assumptions about the value's type.
- Performance: Operations on empty interface values may require type assertions or reflections to determine the underlying type, which can be less efficient than working with statically typed values.

- Readability: Excessive use of the empty interface can make code harder to understand and maintain, as the purpose and expected type of values become less clear.

6.2.4 Type assertions and Type switches

In Go, type assertions and type switches are powerful tools used in conjunction with interfaces to inspect and manipulate the dynamic type of interface variables. These mechanisms allow you to retrieve the concrete type of an item held by an interface or perform different actions based on that type. Understanding how to effectively use type assertions and type switches is crucial for working with interfaces and for writing flexible, type-safe code.

Type Assertions

A type assertion provides a way to access the dynamic value of an interface variable, asserting that the variable holds a value of a specific type. This allows you to retrieve the concrete value from an interface variable.

Syntax

The syntax for a type assertion is:

```
value, ok := interfaceVariable.(Type)
```

Here,

- interfaceVariable is the variable whose dynamic type you want to assert.
- Type is the concrete type you believe interfaceVariable holds.
- value is the variable that will hold the concrete value if the assertion is successful.
- ok is a boolean value that indicates whether the type assertion succeeded.

Example

```
package main
import "fmt"
func main() {
    var i interface{} = "Go"
    // Successful type assertion
    s, ok := i.(string)
    fmt.Println(s, ok) // Output: Go true
    // Failed type assertion
    f, ok := i.(float64)
    fmt.Println(f, ok) // Output: 0 false
}
```

Type Switches

A type switch allows you to perform different actions based on the dynamic type of an interface variable. It's like a regular switch statement but each case specifies types, not values.

Syntax

The syntax for a type switch is:

```
switch v := interfaceVariable.(type) {
case Type1:
    // v has type Type1 here
case Type2:
    // v has type Type2 here
default:
```

```
  // No type match; v has the same type as interfaceVariable
}
```

Here,

- interfaceVariable is the variable whose dynamic type you're switching on.
- v is the variable that holds the concrete value of interfaceVariable in each case block.

Example

```
package main
import "fmt"
func do(i interface{}) {
  switch v := i.(type) {
  case int:
    fmt.Printf("Twice %v is %v\n", v, v*2)
  case string:
    fmt.Printf("%q is %v characters long\n", v, len(v))
  default:
    fmt.Printf("I don't know about type %T!\n", v)
  }
}
func main() {
  do(21)
  do("hello")
  do(true)
```

```
}
```

This example demonstrates how a type switch can be used to handle different types stored in interface variables, performing type-specific actions for each case.

Chapter 7
Concurrency

Concurrency is crucial for building applications that are efficient, responsive, and capable of handling multiple tasks simultaneously. By structuring programs to execute multiple operations in overlapping time periods, concurrency allows a single processor to manage multiple tasks by interleaving their execution or enables multiple processors to handle tasks concurrently. This capability is especially critical in the development of scalable web services, real-time processing systems, and applications that need to maintain high responsiveness under heavy loads.

Imagine you are managing a team where each team member is handling a different task. If every member had to wait for the other to finish before starting their task, it would slow down the entire project. Concurrency in programming is like having each team member work in parallel, making the whole process more efficient and much faster.

Let's see one more analogy, imagine you are in a kitchen preparing a large meal with several dishes. Instead of cooking each dish one at a time, which would be time-consuming and inefficient, you decide to prepare multiple dishes simultaneously. While the roast is in the oven, you could be chopping vegetables for a salad, and while the salad rests in the fridge, you might start boiling pasta for another dish. Each of these cooking tasks overlaps in time but does not interfere with one another.

Concurrency in programming works in a similar way. Instead of executing tasks sequentially which could lead to delays—especially when one task is waiting for resources or data to become available—it allows multiple processes to be executed in parallel. This overlapping execution maximizes the utilization of available computing resources, speeding up the overall process and enhancing the performance of the application.

In this chapter, we'll explore how Go's built-in support for concurrency enables developers to design more efficient and effective programs. We'll delve into goroutines and channels, Go's primary tools for implementing concurrent operations, which help in managing state, synchronizing processes, and ensuring that all tasks are carried out smoothly and efficiently. Through practical examples and detailed explanations, you'll learn how to harness the power of concurrency to build superior software that stands up to the demands of large-scale, real-time applications. This understanding will empower you to write Go programs that maximize resources and performance, making the best use of the system's capabilities to achieve faster and more reliable results.

Definition and Importance of Concurrency

Concurrency refers to the ability of a program to manage multiple tasks at once. Unlike sequential execution, where tasks are completed one after another, concurrent programs can start, execute, and complete tasks in overlapping time periods. This doesn't necessarily mean tasks are executed at the same instant; rather, it involves managing multiple states of execution at the same time within a single application. The importance of concurrency lies in its potential to significantly improve the throughput and scalability of an application, enhance resource utilization, and improve the overall user experience by making applications more responsive.

Overview of Go's Concurrency Model

Go's approach to concurrency is one of its standout features, designed around two primary constructs: goroutines and channels. This model is often summarized by the phrase "Do not communicate by sharing memory; instead, share memory by communicating."

- Goroutines: A goroutine is a lightweight thread managed by the Go runtime. Goroutines are used to perform concurrent tasks and are much more efficient than traditional threads in terms of memory overhead and setup/teardown time. The ease of spawning thousands or even millions of goroutines allows Go applications to handle a vast number of concurrent operations.
- Channels: Channels are the conduits through which goroutines communicate. They allow the safe passage of data between goroutines without the need for explicit locks or condition variables. Channels ensure that data races are avoided by design, facilitating patterns of communication that keep data synchronized as it flows between different parts of a program.

Distinction from Parallelism

While concurrency and parallelism are related concepts, they are not the same. Concurrency is about dealing with multiple tasks at once, which might or might not run in parallel. Parallelism, on the other hand, specifically refers to the simultaneous execution of multiple computations. It's a subset of concurrency that involves executing multiple operations at the same time, typically on multi-core processors.

7.1 Goroutines

Goroutines are a core feature of Go's approach to concurrency and parallelism, enabling developers to create efficient and scalable applications. They represent independently executing functions or methods, essentially lightweight threads managed by the Go runtime rather than the operating system. This distinction from traditional threads offers significant advantages in terms of resource consumption, ease of use, and performance.

Goroutines are much lighter than traditional threads. A key difference lies in how they are scheduled and managed. Traditional threads are managed by the operating system with a fixed stack size (often quite large), leading to significant overhead, especially when handling thousands of concurrent connections or tasks. In contrast, goroutines are managed by the Go runtime and start with a small stack that grows and shrinks as needed. This lightweight nature allows a single Go application to spawn thousands or even millions of goroutines efficiently.

Moreover, the Go runtime multiplexes goroutines onto a smaller number of OS threads, allowing for concurrent execution without the overhead associated with having a corresponding system thread for each task. This model facilitates easier concurrent programming, as it abstracts away many complexities associated with thread management, like blocking I/O operations or system-level context switching.

Syntax

Creating a goroutine in Go is remarkably simple, using the go keyword followed by the function or method call:

```
go functionName(params)
```

Here,

- functionName refers to the function to be executed as a goroutine.
- params are the parameters passed to the function, if any.

Example

```
package main
import (
   "fmt"
   "time"
)
func printMessage(message string) {
   fmt.Println(message)
}
func main() {
   go printMessage("Hello, goroutine!")
```

```
    // Wait for the goroutine to finish
    time.Sleep(1 * time.Second)
}
```

In this example, printMessage is executed as a goroutine. The time.Sleep call is a simple way to wait for the goroutine to complete its execution, which is necessary here because the main program exits as soon as it completes, potentially before the goroutine finishes.

Common Patterns and Practices for Managing Goroutines

- Synchronization: Use channels or synchronization primitives from the sync package (like WaitGroup) to manage the execution and coordination of goroutines.
- Error Handling: Collect errors from goroutines via channels or other synchronization mechanisms to ensure robust error handling.
- Avoiding Goroutine Leaks: Always make sure goroutines can exit, either by completing their task or through explicit cancellation signals, to avoid leaks.
- Limiting Goroutine Numbers: For tasks that can spawn an excessive number of goroutines (e.g., processing input from users or files), use patterns like worker pools to limit the number of concurrent goroutines and manage resources effectively.

7.2 Channels

Channels are a powerful feature in Go's concurrency model, serving as conduits through which goroutines communicate. By providing a way for goroutines to synchronize and exchange data, channels enable safe and elegant solutions to complex concurrent programming challenges. Understanding channels and their usage patterns is crucial for writing effective Go programs.

A channel in Go is a typed conduit that allows you to send and receive values between goroutines. Channels ensure that data exchanges are synchronized: a send operation on a channel will block until another goroutine is ready to receive the data, and vice versa. This synchronization mechanism prevents race conditions and ensures that data is safely shared across goroutines, making channels integral to Go's approach to concurrent programming. Channels can be unbuffered or buffered, and they are created using the make function:

Unbuffered Channel

A channel that does not have any capacity to hold messages before they are received. It ensures that a send operation will block until the message is received.

```
ch := make(chan int) // Create an unbuffered channel of int
```

Buffered Channel

A channel with a specified capacity that can hold a certain number of messages before blocking is needed. Send operations block only when the buffer is full.

```
ch := make(chan int, 5) // Create a buffered channel with a capacity of 5
```

Sending a Value

Use the channel operator <- to send a value to a channel.

```
ch <- value // Send value to channel ch
```

Receiving a Value

Use the same channel operator to receive a value from a channel.

```
value := <-ch // Receive a value from channel ch
```

Example

```
package main
import "fmt"
func main() {
```

```
    messages := make(chan string)
    go func() { messages <- "ping" }() // Send "ping" to the messages channel
    msg := <-messages // Receive the message from the channel
    fmt.Println(msg)  // Output: ping
}
```

This example demonstrates a goroutine sending a message "ping" through the messages channel, which is then received and printed by the main function.

Channel Patterns

- **Fan-in:** Combining multiple results into one channel.

```
func fanIn(input1, input2 <-chan string) <-chan string
{
    c := make(chan string)
    go func() { for { c <- <-input1 } }()
    go func() { for { c <- <-input2 } }()
    return c
}
```

- **Fan-out:** Distributing tasks among multiple goroutines for parallel processing.
- **Time-outs:** Using the select statement with time.After to prevent blocking forever.

```
select {
case msg := <-ch:
```

```
    fmt.Println(msg)
case <-time.After(1 * time.Second):
    fmt.Println("Timed out")
}
```

7.3 Mutexes and Synchronization

Concurrent programming in Go provides powerful tools for building efficient and scalable applications. However, managing access to shared resources among concurrent goroutines introduces the risk of race conditions, where the outcome depends on the non-deterministic ordering of goroutines' execution. Synchronization is crucial to ensure safe access to these shared resources, preventing race conditions and ensuring data integrity.

7.3.1 Overview of Race Conditions

A race condition occurs when two or more goroutines access a shared resource concurrently, and at least one access is a write. Without proper synchronization, the final state of the resource can be unpredictable, leading to bugs that are challenging to reproduce and fix. The need for synchronization arises to enforce an order on the operations performed on shared resources, making the program's behavior deterministic and safe.

7.3.2 Introduction to Mutexes

The sync package in Go provides mutexes to help synchronize access to shared resources. A mutex (mutual exclusion lock) ensures that only one goroutine can access a resource at a time. When a goroutine acquires a mutex lock, any other goroutine attempting to acquire the same lock is blocked until the lock is released.

sync.Mutex

Provides a basic lock mechanism to protect shared resources. A goroutine can lock a Mutex before accessing a shared resource and unlock it afterward.

```
var mu sync.Mutex
```

Example

Protecting a Shared Resource with sync.Mutex

```
package main
import (
    "fmt"
    "sync"
)
var (
    mu  sync.Mutex // A mutex to protect the counter
    counter int
)
func increment() {
    mu.Lock() // Lock the mutex before accessing `counter`
    defer mu.Unlock() // Ensure the mutex is unlocked when the function returns
    counter++ // Safely increment the counter
}
func main() {
    var wg sync.WaitGroup
    for i := 0; i < 1000; i++ {
        wg.Add(1)
        go func() {
            defer wg.Done()
```

```
        increment()
    }()
  }
  wg.Wait() // Wait for all goroutines to finish
  fmt.Println("Counter:", counter) // Output: Counter: 1000
}
```

sync.RWMutex

A read-write mutex that allows multiple goroutines to read a shared resource simultaneously but requires exclusive access for writing. This is useful when a resource is read frequently but written infrequently.

```
var rwmu sync.RWMutex
```

Example

```
package main
import (
  "fmt"
  "sync"
  "time"
)
var (
  rwmu sync.RWMutex
  data int
)
```

```
func read(id int) {
    rwmu.RLock() // Lock for reading
    defer rwmu.RUnlock() // Unlock when done reading
    fmt.Printf("Goroutine %d reads %d\n", id, data)
}
func write(value int) {
    rwmu.Lock() // Lock for writing
    defer rwmu.Unlock() // Unlock when done writing
    data = value
    fmt.Println("Write", value)
}
func main() {
    var wg sync.WaitGroup
    wg.Add(1)
    go func() { defer wg.Done(); write(42) }()
    time.Sleep(100 * time.Millisecond) // Ensure write happens first for example
clarity
    for i := 1; i <= 3; i++ {
        wg.Add(1)
        go func(id int) {
            defer wg.Done()
            read(id)
        }(i)
    }
    wg.Wait()
```

```
}
```

This example demonstrates the use of sync.RWMutex for a scenario where data is written once but read multiple times concurrently. RWMutex allows multiple readers to access the data simultaneously without locking each other out, providing efficient synchronization for read-heavy use cases.

Alternatives to Mutexes for Synchronization

- Channels: Go's channels can be used for synchronization by signaling between goroutines. They are especially useful for coordinating state changes or passing data safely between goroutines.
- Atomic Operations: The sync/atomic package provides low-level atomic operations for synchronizing access to integers and pointers. Atomic operations can be used for counters or flags that are accessed concurrently.

7.4 Best Practices

Concurrent programming in Go offers powerful primitives to build complex concurrent applications efficiently. However, it also introduces challenges such as race conditions, deadlocks, and managing state across multiple goroutines. Adopting best practices can help mitigate these challenges, leading to robust, maintainable, and deadlock-free applications.

Guidelines for Designing Concurrent Programs

- Share Memory by Communicating: Go encourages sharing memory between goroutines by communicating over channels, rather than by sharing variables directly. This approach minimizes shared state and reduces the risk of race conditions, making it easier to reason about the program's behavior.
- Start with a Simple Design: Begin with the simplest concurrency model that could possibly work. Use goroutines and channels sparingly at first. Optimize and add more concurrency only when necessary and when you have a clear understanding of the bottlenecks in your program.

- Make Zero Values Useful: Design your types so that their zero values are safe to use in concurrent contexts. This practice simplifies initialization and reduces the chance of nil pointer dereferences.

Tips for Avoiding Common Pitfalls

- Prevent Race Conditions: Use Go's -race detector during development to identify and fix race conditions. Always access shared variables using channels or synchronization primitives like mutexes. Prefer immutable data structures or deep copying data that needs to be shared across goroutines.
- Avoid Deadlocks: Deadlocks can occur when goroutines wait on each other in a cycle. To avoid deadlocks, ensure that goroutines don't hold locks while waiting for another goroutine. Be cautious with the order of acquiring locks and design your channel communications to prevent goroutines from waiting indefinitely.
- Limit the Number of Goroutines: While goroutines are lightweight, having too many can still lead to high memory usage and scheduling overhead. Use worker pools or semaphores to limit the number of concurrent goroutines, especially when they perform I/O operations or other blocking tasks.

Strategies for Effective Debugging and Testing

- Leverage sync.WaitGroup for Synchronization: Use sync.WaitGroup to wait for groups of goroutines to complete. This is particularly useful in tests where you need to wait for concurrent operations to finish before making assertions.
- Use Buffered Channels as Semaphores: Buffered channels can act as semaphores, allowing you to control the number of goroutines that can perform a certain task at the same time. This pattern is useful for throttling and controlling resource usage.
- Write Deterministic Concurrent Tests: Concurrency can introduce non-determinism in tests. Where possible, design your tests to be deterministic by controlling the execution order of goroutines or by using mock objects and channels to synchronize operations.

- Embrace Concurrency Patterns: Familiarize yourself with common concurrency patterns in Go, such as fan-in, fan-out, worker pools, and rate limiting. These patterns can help solve typical concurrency problems more effectively.
- Regularly Review and Refactor: Concurrent code can be more difficult to understand and maintain than sequential code. Regularly review and refactor your concurrent code to simplify and improve readability. Consider splitting complex functions into smaller, more manageable pieces.

7.5 Advanced Concurrency

As developers grow more comfortable with Go's concurrency primitives, they can explore advanced patterns and techniques that solve complex problems more efficiently. This section delves into sophisticated concurrency patterns, the context package for managing goroutine lifecycles, and the select statement for multiplexing channel operations.

Advanced Concurrency Patterns and Techniques

1. Worker Pools: A worker pool is a concurrency pattern that creates a fixed number of goroutines to perform and manage tasks concurrently. This pattern is effective for controlling resource utilization, managing task execution, and throttling CPU or I/O intensive operations.

```
type Job func()
func startWorkerPool(workerCount int, jobs <-chan Job) {
  for i := 0; i < workerCount; i++ {
    go func() {
      for job := range jobs {
        job()
      }
    }()
```

```
    }
}
```

2. Rate Limiting: Rate limiting controls the rate at which tasks are executed. This can be implemented using time.Ticker or a channel-based approach to ensure that your program does not exceed a rate limit.

```
import "time"
func rateLimiter(requests <-chan int, rate time.Duration) {
    ticker := time.NewTicker(rate)
    defer ticker.Stop()

    for req := range requests {
        <-ticker.C
        // Process request
    }
}
```

Overview of the context Package

The context package provides a means to carry deadlines, cancellation signals, and other request-scoped values across API boundaries and between goroutines. It's particularly useful for managing the lifecycles of goroutines and canceling long-running operations that have become unnecessary or that have exceeded their time limits.

```
func operation(ctx context.Context) {
    select {
```

```
    case <-time.After(5 * time.Second): // Simulate work
        fmt.Println("Operation completed")
    case <-ctx.Done():
        fmt.Println("Operation cancelled")
    }
}
ctx, cancel := context.WithCancel(context.Background())
go operation(ctx)
// Cancel the operation after a timeout or when no longer needed
cancel()
```

Exploration of the select Statement

The select statement in Go allows a goroutine to wait on multiple communication operations, enabling you to handle several channel operations simultaneously. It's akin to a switch statement but for channels.

```
select {
case msg1 := <-channel1:
    fmt.Println("Received from channel1:", msg1)
case msg2 := <-channel2:
    fmt.Println("Received from channel2:", msg2)
case <-time.After(1 * time.Minute):
    fmt.Println("Timeout")
}
```

Chapter 8
Error Handling

This chapter delves into Go's pragmatic approach to error handling, illustrating how it diverges from the more common exception-based models found in many other programming languages. In Go, an error is treated as a value, any type that can describe an abnormal state. The language defines an error as an interface, which is the conventional means for representing an error condition, with the nil value representing the absence of an error.

```
type error interface {
   Error() string
}
```

This design encourages Go programmers to handle errors as part of the normal flow of a program, integrating error management directly into the logic without disrupting the execution with exceptions. This method is akin to a seasoned navigator who plans routes that account for potential obstacles and detours rather than merely reacting to them unexpectedly. By doing so, it allows for more robust and manageable code, as handling what might go wrong is considered just as important as handling what should go right.

In this chapter, we will explore effective patterns for managing errors in your Go programs. You'll learn how to leverage Go's error interface to create clear and predictable error handling pathways in your code. This approach not only simplifies debugging but also enhances the reliability of your applications by making them resilient and adaptive to unexpected conditions. Through practical examples and detailed explanations, you'll gain a thorough understanding of how to implement these strategies to maintain smooth and efficient program execution, even when faced with runtime anomalies.

8.1 Understanding the error Type

The error type's significance lies in its ability to make error handling explicit and integrated into the normal control flow of a program. Functions that can fail return an error value as part of their return signature. This forces calling code to handle errors immediately or propagate them up the call stack, making the presence of errors obvious and unavoidable.

```
func readFile(path string) ([]byte, error) {
   data, err := ioutil.ReadFile(path)
   if err != nil {
      return nil, err // Propagating error up the call stack
   }
   return data, nil
}
```

Comparing Go's Explicit Error Handling with Exception-Based Handling

Unlike languages that use exceptions for error handling, Go's model does not unwind the stack automatically on an error. Instead, errors are values returned by functions that must be checked by the caller. This model encourages developers to think carefully about error handling, leading to more predictable and debuggable code.

8.2 Techniques for Propagating Errors

Unlike exception-based languages where errors (or exceptions) bubble up automatically, Go requires developers to explicitly pass errors back to the caller. This section explores techniques for propagating errors effectively in Go, including best practices for maintaining clarity and usefulness of the error information as it moves through layers of an application.

8.2.1 Returning Errors

The most straightforward technique for error propagation in Go is to return errors from functions to their callers. This involves including an error value in the return signature of any function that might encounter an error.

```
func readFile(filename string) ([]byte, error) {
    data, err := ioutil.ReadFile(filename)
    if err != nil {
        // Directly returning an error to the caller
        return nil, err
    }
    return data, nil
}
```

In this pattern, when an error occurs, the function returns nil for its intended result and the error as it is. The caller is then responsible for checking this error and deciding how to handle it.

8.2.2 Wrapping Errors

As errors propagate up the call stack, it can be helpful to add context to the original error, making it easier to understand where and why the failure occurred. Go 1.13 introduced support for wrapping errors using the %w verb with fmt.Errorf, which allows errors to be unwrapped later.

```
func processFile(filename string) error {
    _, err := readFilc(filename)
    if err != nil {
        // Wrapping the error with additional context
        return fmt.Errorf("processFile failed for %s: %w", filename, err)
    }
    // Further processing...
    return nil
}
```

Wrapped errors retain the underlying original error, which can be examined using functions like errors.Is and errors.As to check for specific error types or conditions.

8.2.3 Custom Error Types

For more complex error handling scenarios, you might define custom error types. These types can carry additional information about the error, beyond just a text message. Custom error types are particularly useful for distinguishing between different error conditions programmatically.

```
type NotFoundError struct {
    Filename string
}
func (e *NotFoundError) Error() string {
    return fmt.Sprintf("file %s not found", e.Filename)
}
func openConfigFile(filename string) (*os.File, error) {
    _, err := os.Stat(filename)
    if os.IsNotExist(crr) {
```

```
        // Returning a custom error
        return nil, &NotFoundError{Filename: filename}
    } else if err != nil {
        return nil, fmt.Errorf("unable to stat the file %s: %w", filename, err)
    }
    return os.Open(filename)
}
```

Chapter 9
Working with Files and I/O

Interacting with the file system is a fundamental requirement for many software applications, from logging data and reading configuration files to processing user-generated content. Go offers robust support for these file operations through its standard library, equipping developers with powerful tools to efficiently manage input and output (I/O) tasks.

Imagine you are organizing a digital library where books, user reviews, and ratings are stored and retrieved frequently. In this scenario, efficient file handling not only speeds up data access but also ensures data integrity and security. Similarly, in a Go application, mastering file operations and I/O handling means your programs can interact seamlessly with the file system, making them capable of managing data with the precision and reliability required in professional environments.

In this chapter, we will delve into how you can leverage Go's file handling packages to create, read, update, and delete files. You'll learn best practices for efficient I/O operations that can help you build faster and more responsive applications. Through practical examples and in-depth explanations, we'll explore the nuances of file handling that are pivotal for any developer looking to harness the full potential of Go in real-world applications.

9.1 Basic File Operations

9.1.1 Opening and Closing Files

The os package in Go is used for interacting with the operating system, including performing file operations. To open a file, you use the os.Open function, which returns a file handle and an error.

```
file, err := os.Open("file.txt")
if err != nil {
    log.Fatal(err)
}
defer file.Close()
```

To create a new file or truncate an existing file, use os.Create:

```
file, err := os.Create("file.txt")
if err != nil {
    log.Fatal(err)
}
defer file.Close()
```

9.1.2 Reading from files

The io.Reader interface is a fundamental part of Go's I/O ecosystem, representing the read end of a stream. The os.File type implements this interface, allowing you to read from files easily.

```
buffer := make([]byte, 100)
n, err := file.Read(buffer)
```

```
if err != nil {
    log.Fatal(err)
}
fmt.Printf("Read %d bytes: %s\n", n, string(buffer[:n]))
```

For more efficient reading, especially of large files, you can use a bufio.Reader:

```
reader := bufio.NewReader(file)
line, err := reader.ReadString('\n')
if err != nil {
    log.Fatal(err)
}
fmt.Println(line)
```

9.1.3 Writing to files

Similarly, the io.Writer interface allows you to write data to a destination, such as a file. The os.File type also implements the io.Writer interface.

```
n, err := file.Write([]byte("Hello, Go!\n"))
if err != nil {
    log.Fatal(err)
}
fmt.Printf("Wrote %d bytes\n", n)
```

For buffered writing, which can improve performance by reducing the number of write system calls, use bufio.Writer:

```
writer := bufio.NewWriter(file)
n, err := writer.WriteString("Buffered Hello, Go!\n")
if err != nil {
   log.Fatal(err)
}
fmt.Printf("Buffered write of %d bytes\n", n)
writer.Flush() // Don't forget to flush!
```

9.2 Different File Formats

9.2.1 JSON File

The encoding/json package provides tools for encoding structs to JSON and decoding JSON data to structs.

```
type Person struct {
   Name string `json:"name"`
   Age  int    `json:"age"`
}
// Decoding JSON to struct
var p Person
err := json.Unmarshal([]byte(`{"name":"John", "age":30}`), &p)
if err != nil {
   log.Fatal(err)
}
fmt.Printf("%+v\n", p)
```

```
// Encoding struct to JSON
b, err := json.Marshal(p)
if err != nil {
    log.Fatal(err)
}
fmt.Println(string(b))
```

9.2.3 CSV File

The encoding/csv package supports reading and writing CSV files, offering flexibility to work with comma-separated data.

```
// Writing to a CSV file
records := [][]string{
    {"Name", "City", "Age"},
    {"John Doe", "New York", "30"},
    {"Jane Doe", "Chicago", "28"},
}
file, err := os.Create("data.csv")
if err != nil {
    log.Fatal(err)
}
defer file.Close()
writer := csv.NewWriter(file)
writer.WriteAll(records) // Writes multiple records at once
if err := writer.Error(); err != nil {
```

```
    log.Fatal(err)
}
// Reading from a CSV file
file, err = os.Open("data.csv")
if err != nil {
    log.Fatal(err)
}
defer file.Close()
reader := csv.NewReader(file)
records, err = reader.ReadAll() // Reads all records at once
if err != nil {
    log.Fatal(err)
}
fmt.Println(records)
```

Chapter 10
Testing and Debugging

Testing and debugging are critical components of the software development lifecycle, crucial for ensuring the reliability, maintainability, and performance of applications. These processes not only help in identifying and fixing bugs but also play a significant role in verifying that the software meets its requirements and performs efficiently under different conditions.

Go provides a comprehensive suite of tools and methodologies designed to support effective testing, debugging, and performance optimization. With built-in support for unit testing through the testing package, developers can easily create and run tests to check individual components of their code. For more complex test scenarios, Go offers benchmarks and profiling tools that help identify performance bottlenecks and optimize code efficiency.

Imagine you're building a bridge. Before opening it to the public, it's essential to test its strength under various loads and inspect it thoroughly to ensure safety and functionality. Similarly, in software development, testing and debugging are like these safety checks, ensuring that your application performs reliably under expected and unexpected conditions.

This chapter explores the strategies and best practices for effective testing and debugging in Go. You'll learn how to use Go's testing framework to write comprehensive test cases, utilize debugging tools to trace and resolve issues, and apply performance optimizations to enhance your application's speed and responsiveness. Through practical examples and detailed guidance, we'll delve into how these practices can be integrated into your development workflow to build robust, efficient, and error-free software.

10.1 Writing Unit Tests

Go's standard library includes a powerful testing package, testing, which integrates seamlessly with the Go toolchain. This package provides the necessary tools to write and execute automated tests for your Go code. Tests in Go are written in Go files, typically alongside the code they test, and the file names end with _test.go.

To run tests, you use the go test command, which automatically locates files named *_test.go, executes them, and reports the results. The simplicity and accessibility of Go's testing framework encourage a test-driven development (TDD) approach.

10.1.1 Writing Test Functions

Test functions in Go are straightforward to write. They must:

- Be in a file that ends with _test.go.
- Start with the word Test.
- Take a single argument of type *testing.T

The *testing.T type provides methods for reporting test failures and logging additional information.

Example

```
package mypackage
import "testing"
func TestSum(t *testing.T) {
    total := Sum(5, 5)
    if total != 10 {
        t.Errorf("Sum was incorrect, got: %d, want: %d.", total, 10)
```

```
    }
}
```

In this example, TestSum is a test function that checks whether the Sum function returns the expected result. If the test fails, t.Errorf is called to report the failure.

10.1.2 Table-Driven Tests

Table-driven tests are a powerful way to test multiple scenarios using the same test logic. They are written as a single test function iterating over a slice of structs, each representing a test case.

Example

```
package mypackage
import "testing"
func TestSumTableDriven(t *testing.T) {
    var tests = []struct {
        a, b int
        want int
    }{
        {1, 2, 3},
        {4, 5, 9},
        {-1, 1, 0},
    }
    for _, tt := range tests {
        testname := fmt.Sprintf("%d+%d", tt.a, tt.b)
        t.Run(testname, func(t *testing.T) {
```

```
        ans := Sum(tt.a, tt.b)
        if ans != tt.want {
            t.Errorf("got %d, want %d", ans, tt.want)
        }
    })
  }
}
```

This example demonstrates a table-driven test for the Sum function, checking multiple input combinations within the same test function.

10.1.3 Mocking and Interfaces

Mocking is a technique used to simulate the behavior of real objects. In Go, interfaces are instrumental in creating mocks. By designing your code around interfaces, you can easily swap out real implementations with mocks during testing.

Example

Assuming you have an interface DataFetcher with a method FetchData, you can create a mock implementation of this interface for testing:

```
type MockFetcher struct {
    DataToReturn string
}
func (m *MockFetcher) FetchData() string {
    return m.DataToReturn
}
// In your test
func TestSomethingThatUsesFetcher(t *testing.T) {
```

```
    mockFetcher := &MockFetcher{DataToReturn: "mock data"}
    // Use mockFetcher in place of a real fetcher
}
```

10.1.4 Testing Concurrency

Testing concurrent code requires considering the synchronization between goroutines to ensure consistent test outcomes.

Example

```
func TestMyConcurrentFunc(t *testing.T) {
    result := make(chan int)
    go MyConcurrentFunc(result)
    r := <-result
    if r != expected {
        t.Errorf("Expected %d, got %d", expected, r)
    }
}
```

10.2 Debugging Techniques

Debugging is an essential part of the development process, allowing developers to identify and resolve issues in their code. Go provides several built-in tools and supports various techniques to facilitate effective debugging. Understanding these tools and techniques can significantly enhance your ability to swiftly diagnose and fix problems in Go applications.

10.2.1 Leveraging Go's Tooling

go vet: The go vet command examines Go source code and reports suspicious constructs, such as potential bugs or structural problems. It doesn't execute the code but statically analyzes it to find issues that may not be syntactic errors but could lead to bugs.

```
go vet ./...
```

golint: golint is a linter for Go source code. It suggests style and coding standards recommendations. While not directly a debugging tool, following golint suggestions can lead to cleaner code that's less prone to bugs.

```
golint ./...
```

Delve (dlv): Delve is a debugger for the Go programming language, offering a command-line interface and features for setting breakpoints, stepping through code, inspecting variables, and more. It allows for in-depth examination of running programs.

```
dlv debug ./path/to/your/package
```

10.2.2 Using Print Statements

One of the simplest yet effective debugging techniques is inserting fmt.Println statements into your code. This method allows you to print the values of variables, execution flow indicators, or custom messages at runtime to the console.

```
func calculateSum(numbers []int) int {
  sum := 0
  for _, num := range numbers {
    fmt.Println("Adding:", num) // Debug print
    sum += num
```

```
    }
    fmt.Println("Total Sum:", sum) // Debug print
    return sum
}
```

Print debugging is most suited for quick checks or when you need immediate feedback without setting up a debugger session. However, it can become cumbersome for more complex debugging scenarios and may clutter your codebase if not removed afterward.

10.2.3 Visual Debuggers and IDEs

Visual debuggers and IDE integrations offer a more sophisticated approach to debugging by providing graphical interfaces for setting breakpoints, stepping through code, inspecting variables, and viewing call stacks. Popular Go IDEs like Visual Studio Code, GoLand, and Atom have built-in support or plugins available for debugging Go code.

Using Visual Studio Code: After setting up the Go extension, you can start debugging by setting breakpoints in the editor and launching a debug session using the built-in debug controls. The interface allows you to step through code, inspect variables, and evaluate expressions on the fly.

10.2.4 Analyzing Stack Traces

When a Go program panics, it generates a stack trace, which is a report of the active stack frames at the point of failure. Analyzing stack traces is crucial for identifying the source of runtime errors or panics. For example to read stack trace -

- Start from the bottom of the trace, which shows the origin of the panic or error.
- Follow the trace upwards to understand the sequence of function calls that led to the issue.

- Look for line numbers and file names to pinpoint where in your code the error occurred.
- Pay attention to any deferred functions or goroutines, as they can influence the program's state at the time of the error.

```
panic: runtime error: index out of range [4] with length 4
goroutine 1 [running]:
main.calculateSum(0xc00001c080, 0x4, 0x4, 0x0)
   /path/to/your/file.go:10 +0x1a5
main.main()
   /path/to/your/file.go:20 +0x72
```

In this example, the stack trace shows a panic caused by an "index out of range" error, indicating the file and line number where the panic occurred.

www.ingramcontent.com/pod-product-compliance
Ingram Content Group UK Ltd.
Pitfield, Milton Keynes, MK11 3LW, UK
UKHW060059200726
13845UKWH00061B/1276

9 789367 834633